THE GENERALIZED RIEMANN INTEGRAL

By

ROBERT M. McLEOD

THE

CARUS MATHEMATICAL MONOGRAPHS

Published by

THE MATHEMATICAL ASSOCIATION OF AMERICA

———

THE CARUS MATHEMATICAL MONOGRAPHS are an expression of the desire of Mrs. Mary Hegeler Carus, and of her son, Dr. Edward H. Carus, to contribute to the dissemination of mathematical knowledge by making accessible at nominal cost a series of expository presentations of the best thoughts and keenest researches in pure and applied mathematics. The publication of the first four of these monographs was made possible by a notable gift to the Mathematical Association of America by Mrs. Carus as sole trustee of the Edward C. Hegeler Trust Fund. The sales from these have resulted in the Carus Monograph Fund, and the Mathematical Association has used this as a revolving book fund to publish the succeeding monographs.

The expositions of mathematical subjects which the monographs contain are set forth in a manner comprehensible not only to teachers and students specializing in mathematics, but also to scientific workers in other fields, and especially to the wide circle of thoughtful people who, having a moderate acquaintance with elementary mathematics, wish to extend their knowledge without prolonged and critical study of the mathematical journals and treatises. The scope of this series includes also historical and biographical monographs.

The following monographs have been published:

No. 1. Calculus of Variations, by G. A. BLISS

No. 2. Analytic Functions of a Complex Variable, by D. R. CURTISS

No. 3. Mathematical Statistics, by H. L. RIETZ

No. 4. Projective Geometry, by J. W. YOUNG

No. 5. A History of Mathematics in America before 1900, by D. E. SMITH and JEKUTHIEL GINSBURG (out of print)

No. 6. Fourier Series and Orthogonal Polynomials, by DUNHAM JACKSON

No. 7. Vectors and Matrices, by C. C. MacDUFFEE

No. 8. Rings and Ideals, by N. H. McCOY

No. 9. The Theory of Algebraic Numbers, Second edition, by HARRY POLLARD and HAROLD G. DIAMOND

The Carus Mathematical Monographs

NUMBER TWENTY

THE GENERALIZED
RIEMANN INTEGRAL

By

ROBERT M. McLEOD
Kenyon College

Published and Distributed by

THE MATHEMATICAL ASSOCIATION OF AMERICA

© *1980 by*
The Mathematical Association of America (Incorporated)
Library of Congress Catalog Card Number 80-81043

Complete Set ISBN 0-88385-000-1
Vol. 20 ISBN 0-88385-021-4

Printed in the United States of America

Current printing (last digit):

10 9 8 7 6 5 4 3 2 1

PREFACE

In calculus courses we learn what integrals are and how to use them to compute areas, volumes, work and other quantities which are useful and interesting. The calculus sequence, and frequently the whole of the undergraduate mathematics program, does not reach the most powerful theorems of integration theory. I believe that the generalized Riemann integral can be used to bring the full power of the integral within the reach of many who, up to now, get no glimpse of such results as monotone and dominated convergence theorems. As its name hints, the generalized Riemann integral is defined in terms of Riemann sums. It reaches a higher level of generality because a more general limit process is applied to the Riemann sums than the one familiar from calculus. This limit process is, all the same, a natural one which can be introduced through the problem of approximating the area under a function graph by sums of areas of rectangles. The path from the definition to theorems exhibiting the full power of the integral is direct and short.

I address myself in this book to persons who already have an acquaintance with integrals which they wish to extend and to the teachers of generations of students to come. To the first of these groups, I express the hope that the organization of the work will make it possible for you to extract the principal results without struggling through technical details which you find formidable or extraneous to your purposes. The technical level starts low at the opening of each chapter. Thus you are invited to follow each chapter as far as you wish and then to skip to the beginning of the next. To readers who do wish to see all the details of the arguments, let me say that they are given. It was a virtual necessity to include them. There are no works to refer you to which are generally available and compatible with this one in approach to integration.

I first learned of the generalized Riemann integral from the pioneering work of Ralph Henstock. I am in his debt for the formulation of the basic concept and for many important methods of proof. Nevertheless, my presentation of the subject differs considerably from his. In particular, I chose to use only a part of his technical vocabulary and to supplement the part I selected with terms from E. J. McShane and other terms of my own devising.

I wish to express my appreciation to the members of the Subcommittee on Carus Monographs for their encouragement. I am particularly indebted to D. T. Finkbeiner. His support enabled me to persevere through the years since this writing project began. He and Helene Shapiro also worked through an earlier version of the book and provided helpful comments.

I thank the Department of Mathematical Sciences of New Mexico State University for its generous hospitality during a sabbatical leave year devoted in large part to the writing of the first version of the book. Finally, I thank Jackie Hancock, Joy Krog, and Hope Weir for expert typing.

Gambier, Ohio   ROBERT M. McLEOD
January, 1980

LIST OF SYMBOLS

ix

CONTENTS

INTRODUCTION

The discussion of the definite integral in elementary calculus commonly starts from an area problem. Given a region under a function graph, how can its area be calculated? The sum of the areas of slender rectangles is a fairly natural approximation. A limit of such sums yields the exact area. When this process is stripped to its essentials the Riemann integral of a function over a given interval stands revealed.

Other geometric and physical quantities, such as volume and work, fit easily into the framework supplied by the concept of the Riemann integral. Moreover the link between the integral and the antiderivative is not hard to make. Thus students can be brought quickly to the evaluation of specific integrals in the context of interesting natural problems. These are among the reasons why the Riemann integral gets first attention when the integral concept is needed.

The Riemann integral has limitations, however. It applies only to bounded functions. The definition does not make sense on unbounded intervals either. Moreover a function which possesses a Riemann integral must exhibit a great deal of regularity. The need for regularity means that the convergence theorems for the Riemann integral are severely restricted. That is, the opportunity to integrate the limit of a sequence by calculating the limit of the sequence of integrals is scant.

Improper integrals are an elementary way to allow for the integration of some unbounded functions and for integration over unbounded intervals. A more basic change is needed to improve the convergence theorems. The class of integrable functions must be enlarged dramatically so that the limit of a sequence of functions can more easily possess the properties needed for existence of the integral.

Lebesgue's definition of the integral does escape these limitations. However it is a definition not suited to introductory discussions of integrals. Thus the Lebesgue integral has not supplanted the Riemann integral. Instead the Riemann integral holds sway during the period when the techniques are being taught. After two or three years the Lebesgue integral is introduced along with a more critical look at the logical underpinnings of the subject.

This two-tier arrangement—first the Riemann integral, then the Lebesgue integral—denies to many potential users of integrals some of the most powerful tools in integration theory. Thus there is good reason to seek to bring together the naturalness of the Riemann integral and the power of the Lebesgue integral.

Rather recently it has been found that an innocent-looking change in the limit process used in the Riemann integral yields an integral with the range and power of the Lebesgue integral. In 1957 J. Kurzweil introduced such a definition. He used it only for special purposes in the study of differential equations. Ralph Henstock developed the generalized Riemann integral fully and showed its relation to the Lebesgue integral.

The presentation which follows is meant to show that the generalized Riemann integral can fill the needs met heretofore by the Riemann and Lebesgue integrals. It can be the first integral defined in calculus courses. It can also be the powerful integral which makes light work of problems requiring interchange of integrals and limit operations, e.g., term-by-term integration of series.

It is not necessary to know the Lebesgue integral to understand what follows. Nor is it necessary to be an expert in Riemann integration. The definitions and basic facts for Riemann integrals are recalled wherever they are needed.

Each chapter opens with an introduction which tells in nontechnical language what the chapter contains. Many individual sections are organized in this way. When a major proposition is first presented the emphasis is on clarifying it with discussion and examples. The more intricate proofs are deferred to specially designated sections placed near the end of each chapter. These sections marked with "S" can be skipped on first reading.

Enough detail is given so that the validity of each example and each proposition can be confirmed by those who care to do so. Those who do not should be able to skip passages without going off track.

Some exercises are included in the text of the various sections to provide the reader an opportunity for active engagement with the main ideas. These exercises are an integral part of the text. Solutions are given in the Appendix. More exercises are given at the end of each chapter. These include extensions of the material of the chapter and some related ideas, as well as further practice with the content of the chapter. No solutions are offered for these exercises.

Clarity and ease in the exposition of any mathematical subject require terminology and notation suited to that subject. Insofar as possible standard terms and symbols have been used. However, the differences between the generalized Riemann integral and both the Riemann integral and the Lebesgue integral made it expedient to supply names and symbols not previously used and to adapt some of those already in use to slightly different circumstances. An index of symbols has been supplied to make it easier to recover the significance of symbols when they reappear after their first introduction.

The first three chapters should be read in the order in which they are given. After Chapter 3 any one of Chapters 4, 6, and 7 may be taken up. (It may be necessary to look back for specific items of terminology or notation or a specific proposition.) Chapter 5 requires Chapter 4 as a precursor. Chapter 8 may be begun after the first two sections of Chapter 5 have been read.

The last section of Chapter 8 contains some suggestions about further reading on integration theory for those whose appetites have been whetted.

CHAPTER 1

DEFINITION OF THE GENERALIZED RIEMANN INTEGRAL

The first objective in Chapter 1 is to make plain what change is being made in the Riemann definition and to indicate why it should be beneficial. Then the definition of the generalized Riemann integral is formulated. It is given first for bounded intervals of real numbers and then, in the same language, for unbounded intervals. On the basis of these formulations the fundamental theorem of calculus linking integrals and derivatives is given an appealing form. Multiple integrals are defined, and again it is possible to use the same language as that first adopted for integration on a bounded real interval.

A more detailed description of the contents of Chapter 1 follows.

Section 1.1 begins by recalling the definition of the Riemann integral. Then it turns to the familiar problem of the approximation of the area of the region under a function graph and notes how the approximation could be done more effectively. That discussion leads to the crucial notion of a *gauge*. Several examples are worked out in detail to provide insight into the effective use of gauges. The section closes with the statement of a fact about gauges which is indispensable to the satisfactory use of gauges.

Section 1.2 opens with the context in which the definition will be placed. Then some notation for Riemann sums is introduced. At that point the formal definition of the generalized Riemann integral on a bounded interval of real numbers is brought in. The uniqueness of the integral is asserted. The closing paragraphs examine the relation of the generalized Riemann integral to the Riemann integral. Certain functions are shown to have a generalized Riemann integral but not a Riemann integral.

Section 1.3 goes back to the area approximation problem to get at a proper definition of the integral on unbounded intervals. A way is found to raise the notations and language of the bounded interval definition to the full generality of bounded and unbounded intervals. An example points up a technique for effective estimation of sums over subintervals of an unbounded interval.

Section 1.4 deals with the connection between integrals and derivatives. A review of several previous examples suggests the degree of generality which is achievable in the fundamental theorem when the generalized Riemann integral is used. The formal statement of the theorem is followed by some comments on its uses.

Section 1.5 addresses the question of the place of improper Riemann integrals in the context of the generalized Riemann integral. It turns out that the existence and nonexistence of limits which distinguish between convergence and divergence of improper integrals also serve to distinguish integrability from nonintegrability in the generalized Riemann sense.

Section 1.6 defines multiple integrals in the same words and symbols used for integrals over intervals of real numbers. Both bounded and unbounded multidimensional intervals are included in the definition. The definition of the integral over noninterval sets is given in terms of integration over an interval. The use of iterated integrals in the evaluation of multiple integrals is recalled.

Section 1.7 contains a miniature integral, as it were. The convergence of series is recast as an integration process on functions defined on the positive integers.

Section S1.8 can be passed by on first reading. It contains detailed proofs of some basic facts about the limit process on which the generalized Riemann integral is based.

Section S1.9 opens with some deductions from the definition of the derivative. They are needed in the proof of the fundamental theorem which follows. (They will also be useful later.)

1.1. Selecting Riemann sums. The formation of Riemann sums is the beginning point of the definition of the Riemann integral and the generalized Riemann integral. Suppose a function f is to be integrated over the closed interval $[a, b]$. Form a *division* of $[a, b]$ into nonoverlapping subintervals $[x_{k-1}, x_k]$ by choosing numbers x_k such that

$$a = x_0 < x_1 < x_2 < \cdots < x_n = b.$$

Make a *tagged division* out of this division by choosing a number z_k, called a *tag*, in each interval $[x_{k-1}, x_k]$; thus

$$x_{k-1} \leqslant z_k \leqslant x_k$$

for $k = 1, 2, \ldots, n$. Now the number

$$\sum_{k=1}^{n} f(z_k)(x_k - x_{k-1})$$

is a Riemann sum for f on the interval $[a, b]$.

The general idea is that well-chosen Riemann sums will cluster around a number, and that number will be called the integral. The distinction between the Riemann integral and the generalized Riemann integral lies in the criterion used to select Riemann sums so that they will be close approximations to the integral. The usual procedure focuses on sums coming from divisions in which all subintervals are short. The precise definition follows.

Definition of the Riemann integral. The number A is an integral of f on $[a, b]$ provided that to each positive number ϵ there is a positive δ such that

$$\left| A - \sum_{k=1}^{n} f(z_k)(x_k - x_{k-1}) \right| < \epsilon$$

whenever $x_k - x_{k-1} < \delta$ for $k = 1, 2, \ldots, n$.

The area interpretation of the integral will help to clarify the weakness of this definition and to point the way to an alternative.

Suppose $f(x) \geqslant 0$ for $a \leqslant x \leqslant b$ and let S be the region under the graph of f. Each term $f(z_k)(x_k - x_{k-1})$ is then the area of a rectangle, as shown in Figure 1. Moreover, the Riemann sum is an approximation to the area of S. Pictorial evidence suggests that it is advantageous to make the width of the rectangle, i.e., $x_k - x_{k-1}$, small wherever the graph of f is steep. On the other hand, in an interval where f changes little the rectangle may be wider without serious damage to the accuracy of the approximation. What is needed is a way to specify which intervals need to be narrow and which intervals may be allowed to be wide. It is the local behavior of f which is to be used in this specification.

In forming a Riemann sum it is natural to think of choosing the division first and the tags $z_1, z_2, \ldots, z_n$

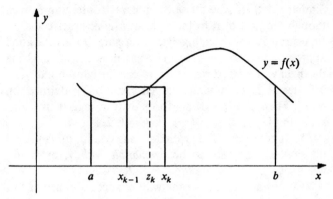

FIG. 1

afterward. But in deciding whether the tagged division has been chosen well, it is better to think of z_k as fixed. Then the question is whether x_{k-1} and x_k are close enough to z_k to make $f(z_k)(x_k - x_{k-1})$ a good approximation to the area of the strip under the graph between the lines $x = x_{k-1}$ and $x = x_k$.

The question of the previous paragraph can be stated without subscripts as follows: Let z be given in $[a, b]$. For what intervals $[u, v]$ containing z is it true that $f(z)(v - u)$ is a close approximation to the area of the strip under the graph between the lines $x = u$ and $x = v$? Naturally, we expect that u can be any number in some small interval to the left of z and v can be any number in some small interval to the right of z. In other words, there will be an open interval $\gamma(z)$ containing z such that $f(z)(v - u)$ is a good area approximation whenever $z \in [u, v]$ and $[u, v] \subseteq \gamma(z)$. Of course $[u, v] \subseteq [a, b]$ also.

When $z = a$ it follows that $u = a$ and the part of $\gamma(a)$ to the left of a is of no importance. Nevertheless it is convenient to let $\gamma(a)$ be an open interval containing a.

Similarly let $\gamma(b)$ be an open interval containing b even though the part of it to the right of b is of no effect.

A strategy for selecting Riemann sums to give a good approximation to the area of S has now emerged. Using the behavior of f at z, assign to z a neighborhood $\gamma(z)$. The result is an interval-valued function γ defined on $[a, b]$. Then consider sums formed from tagged divisions such that $[x_{k-1}, x_k] \subseteq \gamma(z_k)$ for $k = 1, 2, \ldots, n$.

The function γ will be called a *gauge* on $[a, b]$. A tagged division will be said to be *compatible with* γ, or γ-*fine*, when $[x_{k-1}, x_k] \subseteq \gamma(z_k)$ for $k = 1, 2, \ldots, n$.

Two obstacles stand in the way of successful use of this strategy. The strategy is designed to produce a small difference between $f(z_k)(x_k - x_{k-1})$ and the area of the strip bounded by $x = x_{k-1}$ and $x = x_k$. The overall error is the sum of these differences. It is possible that small individual errors could accumulate into a large overall error. The outcome seems even more doubtful when we note that making $\gamma(z)$ a small interval for every z has the unfortunate consequence of requiring a large number of intervals $[x_{k-1}, x_k]$ to cover $[a, b]$. Some examples will help to show that this is not an insuperable obstacle.

Successful avoidance of the first obstacle puts us face to face with the second. When there are arbitrarily small intervals among the values of γ it is not apparent that there will be any γ-fine divisions of $[a, b]$. We will examine this question in individual examples and also give a theorem which shows that there are always γ-fine divisions.

The following examples show how the properties of f dictate the choice of γ. The discontinuities of f and the steepness of the graph of f play the crucial roles.

The first example is a simple one, yet it provides an opportunity to point out a useful consequence of defining a gauge in a special way.

Example 1. Let $f(0) = 3$ and $f(x) = 2$ when $0 < x \leqslant 1$. Define a gauge γ which selects Riemann sums which differ from the area of the region under the graph by at most ϵ.

Solution. The region is a rectangle plus a spike above $(0, 0)$. The spike contributes nothing to the area but it affects all those Riemann sums in which $z_1 = 0$. The Riemann sum $\sum_{k=1}^{n} f(z_k)(x_k - x_{k-1})$ equals 2 when $z_1 > 0$, and it equals $3(x_1 - 0) + 2(1 - x_1)$, or $2 + x_1$, when $z_1 = 0$. It is clear that only $\gamma(0)$ need be made small. The choice $\gamma(0) = (-\epsilon, \epsilon)$ and $\gamma(z) = (-1, 2)$ when $0 < z \leqslant 1$ implies

$$\left| 2 - \sum_{k=1}^{n} f(z_k)(x_k - x_{k-1}) \right| < \epsilon$$

when $[x_{k-1}, x_k] \subseteq \gamma(z_k)$ for $k = 1, 2, \ldots, n$.

The gauge γ which has just been defined responds fully to the requirements stated in Example 1. There is a great deal of freedom in defining a gauge to meet those requirements. It will be instructive to examine an alternative choice of a particular kind.

In Example 1 the function f has special behavior at $x = 0$. Can we define a gauge which will prevent the use of 0 as a tag? Is there another gauge which will force the use of 0 as the tag of the first interval? The first question has a negative answer, but the second has an affirmative answer. Let $\gamma_1(0) = (-\epsilon, \epsilon)$ and $\gamma_1(z) = (0, 2)$ when $0 < z \leqslant 1$. Clearly γ_1, like γ, selects Riemann sums which differ from 2 by less than ϵ. But every γ_1-fine tagged division of $[0, 1]$ has $z_1 = 0$. It cannot be otherwise since $\gamma_1(z)$ contains 0 only when $z = 0$ and z_1 is the tag of an interval $[0, x_1]$ which is a subset of $\gamma_1(z_1)$.

Note also that γ_1 is more selective than γ. That is, every γ_1-fine division is also γ-fine. This follows from the observation that $\gamma_1(z) \subseteq \gamma(z)$ for every z in $[0, 1]$. We say that γ_1 is *stricter than* γ.

The next example also provides an opportunity to force the use of 0 as a tag. The new element it introduces is a method for the control of error accumulation.

Example 2. Let $f(0) = 0$ and $f(x) = 1/\sqrt{x}$ when $0 < x \leqslant 1$. Define a gauge γ on $[0, 1]$ which selects Riemann sums which differ from the area of the region under the graph by at most ϵ.

Solution. The region is an unbounded one, but its area can be found by the improper integral technique. Since $2\sqrt{x}$ is an antiderivative of $1/\sqrt{x}$ when $x > 0$, the area of the region is

$$\lim_{s \to 0} \int_s^1 (1/\sqrt{x})\,dx = \lim_{s \to 0}(2 - 2\sqrt{s}) = 2.$$

Moreover, the area of the strip bounded by $x = u$ and $x = v$ is $2\sqrt{v} - 2\sqrt{u}$. This is true even when $u = 0$.

The challenge here is to analyze the error when $2\sqrt{v} - 2\sqrt{u}$ is approximated by $f(z)(v - u)$ with $u \leqslant z \leqslant v$, i.e., to analyze $2\sqrt{v} - 2\sqrt{u} - f(z)(v - u)$.

We may plan to define $\gamma(z)$ so that $\gamma(z) \subseteq (0, \infty)$ when $0 < z \leqslant 1$. The consequence will be that the first interval $[0, x_1]$ has the tag $z_1 = 0$. Since $f(0) = 0$, the error for the strip between $x = 0$ and $x = x_1$ will be $2\sqrt{x_1}$. To make this error small it is enough to force x_1 to be close to zero by the choice of $\gamma(0)$.

There remains the analysis of the error $2\sqrt{v} - 2\sqrt{u} - (1/\sqrt{z})(v - u)$ when $0 < u \leqslant z \leqslant v$. Since the number of strips to be used is not predictable, the error in each strip must be estimated in such a way that the sum of all the errors can be brought under tight control. The best way to do this is to find a dominant having $(v - u)^2$ as a factor.

Simple algebra suffices. Start with $\sqrt{v} - \sqrt{u} = (v - u)/(\sqrt{v} + \sqrt{u})$. Then factor out $v - u$ and get a common denominator. Replace $\sqrt{v} + \sqrt{u}$ by $\sqrt{z}$ in the denominator. The result is

$$\left| 2\sqrt{v} - 2\sqrt{u} - \frac{1}{\sqrt{z}}(v - u) \right| \leqslant \frac{v - u}{z} |2\sqrt{z} - \sqrt{v} - \sqrt{u}|.$$

Now $|\sqrt{z} - \sqrt{v}| \leqslant (v - z)/\sqrt{z}$ and $|\sqrt{z} - \sqrt{u}| \leqslant (z - u)/\sqrt{z}$. Use these after applying the triangle inequality. The final result is

$$\left| 2\sqrt{v} - 2\sqrt{u} - \frac{1}{\sqrt{z}}(v - u) \right| \leqslant \frac{(v - u)^2}{z\sqrt{z}}.$$

One factor $v - u$ in the majorant $(v - u)^2/(z\sqrt{z})$ will be used to cancel $z\sqrt{z}$ through the choice of $\gamma(z)$. The remaining factor $v - u$ will control the build-up of errors under summation.

Specifically we may set $\gamma(0) = (-\epsilon^2/16, \epsilon^2/16)$ and $\gamma(z) = (z - \delta_z, z + \delta_z)$ with $\delta_z = \epsilon z\sqrt{z}/4$ when $0 < z \leqslant 1$. Note that $0 < z - \delta_z$ since we may assume ϵ is small, say $\epsilon < 1$. Thus $0 \notin \gamma(z)$ when $z > 0$.

These choices of $\gamma(z)$ imply that $|2\sqrt{v} - 2\sqrt{u} - f(z) \cdot (v - u)|$ is less than $\epsilon/2$ when $z = 0$ and less than $(\epsilon/2)(v - u)$ when $0 < u \leqslant z \leqslant v$, provided that $[u, v] \subseteq \gamma(z)$.

Now consider a γ-fine division of $[0, 1]$. The error for the first strip is at most $\epsilon/2$. The sum of the errors for the remaining strips is less than $\sum_{k=2}^{n}(\epsilon/2)(x_k - x_{k-1})$ or

$(\epsilon/2)(1 - x_1)$. Consequently

$$\left| 2 - \sum_{k=1}^{n} f(z_k)(x_k - x_{k-1}) \right| < \epsilon,$$

as required.

The next example presents another technique for the prevention of a dangerous accumulation of error. In it the function f is discontinuous at an infinite sequence of points.

Example 3.   Let $f(1/m) = m$ for every positive integer m. Let $f(x) = 0$ for all other x in $[0, 1]$. Define a gauge γ which selects Riemann sums which differ from the area of the region under the graph by at most ϵ.

Solution.   The region is a unit segment on the x-axis surmounted by a sequence of spikes rising to the left. Thus the area of the region is zero. Our objective is to define γ so that each γ-fine division yields a Riemann sum whose value is less than ϵ.

In the Riemann sum $\sum_{k=1}^{n} f(z_k)(x_k - x_{k-1})$ it is only terms for which $z_k = 1/m$ for some integer m which need concern us. The others are zero. Consequently $\gamma(z)$ can be any open interval containing z when z is not of the form $1/m$, say $\gamma(z) = (-2, 2)$.

When $z = 1/m$ we must choose $\gamma(z)$ so that its length is small enough to cause $x_k - x_{k-1}$ to counteract the factor $f(z_k)$ when $1/m = z_k$ for some k. It is not enough that each term $f(z_k)(x_k - x_{k-1})$ be less than a small constant since the number of non-zero terms is not bounded. To insure a small sum the majorant of $f(z_k)(x_k - x_{k-1})$ must decrease rapidly as m increases, given that $z_k = 1/m$. It is convenient to use the numbers $\epsilon/2^m$ as majorants since

$\sum_{m=1}^{\infty} \epsilon/2^m = \epsilon$. With this objective let $\gamma(1/m) = (1/m - \delta_m, 1/m + \delta_m)$ with $\delta_m = \epsilon/(m2^{m+1})$.

Now consider a γ-fine division of $[0, 1]$. For each m the number $1/m$ may not be a tag or it may be the tag of one interval or of two adjacent intervals. When $1/m = z_k$ for one k, $f(z_k)(x_k - x_{k-1}) < \epsilon/2^m$ since $f(z_k) = m$ and $[x_{k-1}, x_k] \subseteq \gamma(1/m)$. When $1/m = z_{k-1} = z_k$,

$$f(z_{k-1})(x_{k-1} - x_{k-2}) + f(z_k)(x_k - x_{k-1})$$
$$= m(x_k - x_{k-2}) < \epsilon/2^m$$

since x_{k-2} and x_k are in $\gamma(1/m)$. The Riemann sum is a finite sum. Thus it can be rearranged and grouped according to values of m. Then

$$\sum_{k=1}^{n} f(z_k)(x_k - x_{k-1}) \leqslant \sum_{m \in E} \epsilon/2^m$$

where E is the set of integers m such that $1/m$ is a tag. But $\sum_{m \in E} \epsilon/2^m < \sum_{m=1}^{\infty} \epsilon/2^m$ since all the terms on the right are positive. Since the sum of this geometric series is ϵ we have reached the desired conclusion.

The three examples have illustrated ways to define a gauge γ so that each γ-fine division gives rise to a Riemann sum which approximates the area under the graph with small error. Now we ask whether there are γ-fine divisions of $[0, 1]$ in each of these instances.

In Example 1 we chose $\gamma(0) = (-\epsilon, \epsilon)$ and $\gamma(z) = (-1, 2)$ when $0 < z \leqslant 1$. Clearly $[0, 1]$ tagged with any number in $(0, 1]$ is a γ-fine division of $[0, 1]$. Also $[0, \epsilon/2]$ and $[\epsilon/2, 1]$ tagged with 0 and 1 respectively form another γ-fine division. The latter is also γ_1-fine where $\gamma_1(0) = (-\epsilon, \epsilon)$ and $\gamma_1(z) = (0, 2)$ when $0 < z \leqslant 1$.

The gauge γ in Example 2 has the values $\gamma(0) =$

$(-\epsilon^2/16, \epsilon^2/16)$ and $\gamma(z) = (z - \delta_z, z + \delta_z)$ with $\delta_z = \epsilon z \sqrt{z}/4$ when $0 < z \leqslant 1$. Fix x_1 so that $0 < x_1 < \epsilon^2/16$ and h so that $0 < h < \epsilon x_1 \sqrt{x_1}/4$. Select the smallest integer n such that $x_1 + (n - 1)h \geqslant 1$. Let $x_2 = x_1 + h$, $x_3 = x_1 + 2h, \ldots, x_{n-1} = x_1 + (n - 2)h$, and $x_n = 1$. Also set $x_0 = 0$. Let $z_k = x_{k-1}$ for $k = 1, 2, \ldots, n$. The choice of x_1 implies that $[x_0, x_1] \subseteq \gamma(0)$. The choice of h implies that $[x_1, x_2] \subseteq \gamma(x_1)$. Since the length of $\gamma(z)$ is an increasing function of z for $0 < z \leqslant 1$, it is also true that $[x_{k-1}, x_k] \subseteq \gamma(x_{k-1})$ for $k = 3, 4, \ldots, n$. Thus this division is γ-fine.

In Example 3 there is a trivial choice of a γ-fine division. Simply tag $[0, 1]$ with $2/3$.

There was total freedom in the choice of $\gamma(z)$ when z is not the reciprocal of an integer in Example 3. The following exercise presents an alternative which makes the definition of a γ-fine division more challenging.

Exercise 1. Let $\gamma(1/m) = (1/m - \delta_m, 1/m + \delta_m)$ with $\delta_m = \epsilon/(m2^{m+1})$ for $m = 1, 2, 3, \ldots$. When $1/(m + 1) < z < 1/m$ set $\gamma(z) = (1/(m + 1), 1/m)$. Finally set $\gamma(0) = (-\epsilon, \epsilon)$. Construct a γ-fine division of $[0, 1]$.

It would be easy to define gauges even more elaborate than the one in Exercise 1. One may well ask whether there is a gauge γ for which no tagged division is γ-fine. Happily there is not.

COMPATIBILITY THEOREM.   *Let γ be a gauge on $[a, b]$. There is a tagged division of $[a, b]$ which is γ-fine.*

This assertion is closely related to the Heine-Borel theorem. The proofs are very similar and there are many common uses. A proof is given in Section S1.8. Two uses of the Compatibility Theorem outside the theory of integration are proposed in Exercises 16 and 17 at the end of this chapter. Others will come to mind with just a little thought.

1.2. Definition of the generalized Riemann integral.

The groundwork has now been laid for a definition of the integral which differs from the familiar one recalled in Section 1.1 just enough to yield a markedly superior integral. It is appropriate to formulate the definition in as much generality as is natural. At present we propose to consider a function f defined on a closed interval $[a, b]$ in the real numbers $\mathbf{R}$. We may ask what kind of values f must have for our purposes. The essentials are that the Riemann sum be meaningful and that we be able to judge the accuracy of the approximation. Both of these are met when f has values in $\mathbf{R}^q$, the set of q-tuples of real numbers. To measure the accuracy of approximations we shall use the Euclidean length. For x in $\mathbf{R}^q$, with $x = (x_1, x_2, \ldots, x_q)$, $|x|$ will denote $[\sum_{i=1}^{q} x_i^2]^{1/2}$. The case $q = 1$ will be identified with $\mathbf{R}$. For our purposes the case $q = 2$ is indistinguishable from the complex numbers.

It is highly desirable to have a concise and flexible notation for tagged divisions and Riemann sums. For divisions and tagged divisions a script letter, often $\mathcal{D}$, $\mathcal{E}$, or $\mathcal{F}$, will be used. These may have subscripts and superscripts on occasion, too. The definition of tagged division may be rephrased as follows. A *tagged division* $\mathcal{D}$ of $[a, b]$ is a set of ordered pairs, say $\{(z_1, J_1), (z_2, J_2), \ldots, (z_n, J_n)\}$, where $J_1, J_2, \ldots, J_n$ are nonoverlapping closed intervals whose union is $[a, b]$ and $z_k \in J_k$ for $k = 1, 2, \ldots, n$. Very often we need to refer to a single pair (z_k, J_k) in $\mathcal{D}$ but have no particular need for its subscript k. We will be aided if we adopt a notation free of subscripts. We may make a further simplification by using zJ in place of (z, J). Suppose we let $L(J)$ be the length of J, i.e., $L(J) = v - u$ when $J = [u, v]$. Then $f(z)L(J)$ is the term in the Riemann sum corresponding to the tagged interval zJ. To achieve a simple notation for the Riemann sum let us abbreviate $f(z)L(J)$ as $fL(zJ)$ and let $fL(\mathscr{D})$ denote the Riemann sum given by the tagged division $\mathscr{D}$. That is,

$$fL(\mathscr{D}) = \sum_{zJ \in \mathscr{D}} fL(zJ) = \sum_{zJ \in \mathscr{D}} f(z)L(J).$$

Definition of the generalized Riemann integral. Let $f : [a, b] \to \mathbf{R}^q$ be given. An element A of $\mathbf{R}^q$ is *an integral* of f on $[a, b]$ provided the following holds. For each positive ϵ there is a gauge γ such that $|A - fL(\mathscr{D})| < \epsilon$ whenever $\mathscr{D}$ is a γ-fine tagged division of $[a, b]$.

Examples 1, 2, and 3, while cast in the guise of area approximations, actually contain proofs of the existence of the integrals in question. In these examples the area interpretation leads directly to a natural candidate for a value of the integral. Have we, by focusing attention on area, overlooked other numbers which can be approximated arbitrarily closely by properly selected Riemann sums? Fortunately not.

UNIQUENESS THEOREM.   *A given function $f : [a, b] \to \mathbf{R}^q$ has at most one integral on $[a, b]$.*

A proof is given in Section S1.8.

It is not quite obvious that every Riemann integrable function also has an integral in the generalized Riemann sense. The next exercise suggests a way to remove any doubts.

Exercise 2.   (a) Consider tagged divisions $\mathscr{D}$ of $[a, b]$. Let R_δ be the set of all $\mathscr{D}$ such that $L(J) < \delta$ for all zJ in $\mathscr{D}$. Let GR_δ be the set of all $\mathcal{D}$ such that $J \subseteq (z - \delta, z + \delta)$ for all zJ in $\mathcal{D}$. Show that $R_\delta \subseteq GR_\delta \subseteq R_{2\delta}$.

(b) Show that the Riemann integral is a special case of the generalized Riemann integral. Characterize the Riemann integrable functions as a subclass of the functions which are integrable in the generalized Riemann sense.

Since the new notion of integral contains the familiar one as a special case, no conflict can arise from the use of the familiar notations for integrals. Accordingly

$$\int_a^b f(x)dx, \qquad \int_a^b f, \quad \text{or} \quad \int_{[a, b]} f$$

will be used. The choice among them will be dictated by convenience.

One objective in seeking a modification of the Riemann definition was to be able to integrate functions to which the Riemann definition is not applicable. Examples 2 and 3 contain such functions. It is a basic fact about the Riemann integral, as defined in Section 1.1, that only bounded functions are integrable. Neither of the functions in Examples 2 and 3 is bounded. Thus neither is Riemann integrable.

Even within the class of bounded functions the generalized Riemann integral is more widely applicable. The next example may be used to demonstrate this point.

Example 4. Prove integrability and evaluate the integral when f is constant on the complement of a countably infinite subset of $[a, b]$.

Solution. A set C is countably infinite when $C = \{c_1, c_2, c_3, \ldots \}$. Thus we may assume that $f(x) = K$ for all x not in C. (Nothing is claimed about whether $f(x) \neq K$ when $x \in C$. Thus this example includes constant functions.)

The only possible candidate for $\int_a^b f$ is $K(b - a)$ since this is the value of $\int_a^b f$ when f is constant on $[a, b]$. Our objective is to define a gauge γ so that $|K(b - a) - fL(\mathcal{D})| < \epsilon$ for all γ-fine divisions $\mathcal{D}$. Since

$$K(b - a) - fL(\mathcal{D}) = \sum_{zJ \in \mathcal{D}} [K - f(z)]L(J)$$

and $f(z) = K$ when $z \notin C$, the only challenge lies in defining $\gamma(z)$ when $z \in C$. When $z \notin C$ no restriction on $L(J)$ is needed and we may choose $\gamma(z) = (z - 1, z + 1)$.

Each c_n in C is the tag for at most two intervals in a division of $[a, b]$. Keeping this in mind there is a simple choice for $\gamma(c_n)$. It is to choose δ_n so that $|M - f(c_n)|\delta_n \leqslant \epsilon/2^{n+2}$ and set $\gamma(c_n) = (c_n - \delta_n, c_n + \delta_n)$. Then $|M - f(z)|L(J) \leqslant \epsilon/2^{n+1}$ when $z = c_n$ and $J \subseteq \gamma(z)$. Moreover the sum of all terms tagged with c_n is at most $\epsilon/2^n$. Now we group and order the nonzero terms $[K - f(z)]L(J)$ according to the subscript n for which $z = c_n$. Thus $|K(b - a) - fL(\mathcal{D})| < \sum_{n=1}^{\infty} \epsilon/2^n$ when $\mathcal{D}$ is γ-fine. The last sum equals ϵ. Thus the solution is complete.

The set of all reciprocals of positive integers, i.e., the set $\{1, 1/2, 1/3, \dots\}$ of Example 3, is a countably infinite set. A more surprising countably infinite set is the set of all rational numbers between 0 and 1. (We can enumerate them as follows. Put each rational in lowest terms. Begin the list with $1/2, 1/3, 2/3, 1/4, 3/4$. Continue with those having denominator 5, then 6, 7, etc.) Between any two irrational numbers in $(0, 1)$ there is a rational number. Thus the rational numbers are spread throughout $[0, 1]$ so finely that no interval is free of them. In particular they are not isolated from one another. This is a sharp contrast with the set $\{1, 1/2, 1/3, \dots\}$ where each element $1/n$ is contained in an open interval which contains no other reciprocal of a positive integer.

Since there is no limitation on C in Example 4 except that it be countably infinite, we can let C be the set of all rational numbers in $[0, 1]$. The function which is zero on the rationals and one on the irrationals is therefore a special case of Example 4. Its integral exists and equals one. It is a small task (left to the reader) to show that this bounded function is not Riemann integrable.

Not only the function which is zero on the rationals and one on the irrationals but all those bounded functions which are discontinuous at too many points fail to be Riemann integrable. We need not fear a limitation of this type in the generalized Riemann integral since the example cited is discontinuous everywhere on $[0, 1]$ but integrable in the generalized Riemann sense.

Example 4 also shows us that any function which is zero except possibly on a countably infinite set in $[a, b]$ has zero as its integral. Questions related to this observation will be taken up again in Section 4.1 where we consider null functions.

In Examples 2 and 3 we have integrable functions which are not bounded. Their unboundedness occurs only near $x = 0$, however. Much more pathological examples can be obtained from Example 4.

Exercise 3.   Construct a function which is unbounded on every subinterval of $[a, b]$ but which has an integral on $[a, b]$.

1.3. Integration over unbounded intervals. Integrals over unbounded intervals, e.g., $\int_0^\infty f(x)\,dx$, are defined in elementary calculus as limits of integrals over bounded intervals. It is our purpose here to give a definition directly in terms of Riemann sums. Indeed we want the definition to be formally indistinguishable from the one already given for integration over bounded intervals.

The approximation of the area under a curve can again be used to see how to frame a definition. Consider a function f defined on $[a, \infty)$ with positive values. Assume the region under its graph has finite area. How might that area be approximated closely by a sum of areas of rectangles? Again it is helpful to make the rectangles narrow where the graph is steep. A finite number of rectangles cannot cover all of $[a, \infty)$. They should cover a very large interval $[a, t]$. For a sufficiently large t and a well-chosen tagged division of $[a, t]$ the corresponding Riemann sum should approximate the area under the graph closely.

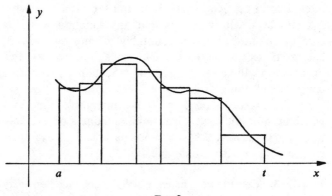

FIG. 2

The kind of definition suggested above can be stated in the language already used for bounded intervals if we enlarge $\mathbf{R}$ by the introduction of points $-\infty$ and ∞. Let $\overline{\mathbf{R}} = \mathbf{R} \cup \{-\infty, \infty\}$. The extended real number set $\overline{\mathbf{R}}$ will be ordered in the natural way, namely, $-\infty < x$ and $x < \infty$ for all x in $\mathbf{R}$ and $-\infty < \infty$. The intervals in $\overline{\mathbf{R}}$ consist of all the intervals in $\mathbf{R}$ along with new open intervals of the form $[-\infty, a)$ and $(a, \infty]$, new closed intervals $[-\infty, a]$ and $[a, \infty]$, and $\overline{\mathbf{R}}$ itself. $\overline{\mathbf{R}}$ is both open and closed, despite the appearance given by the interval notation $[-\infty, \infty]$. An interval in $\overline{\mathbf{R}}$ will be said to be *bounded* when both its endpoints are in $\mathbf{R}$ and *unbounded* when at least one endpoint is $-\infty$ or ∞.

Let $[a, b]$ be a closed interval in $\overline{\mathbf{R}}$. The definitions given in Section 1.1 for a division, a tagged division, and a gauge need no change for an unbounded interval. A Riemann sum cannot be carried over unchanged because there is no length in the ordinary sense for an unbounded interval. The Riemann sum should have a zero term for each unbounded interval in the division of $[a, b]$. One way to guarantee that is simply to extend the definition of L by decreeing that $L(J) = 0$ whenever J is unbounded. With this definition $fL(\mathcal{D})$ is defined as before.

Only an unbounded interval in a division can have $-\infty$ or ∞ as a tag. Thus the value of f at $-\infty$ or ∞ will affect no Riemann sum. Actually a function which is to be integrated on an unbounded interval usually does not come to us with a value assigned at $-\infty$ or ∞. It is immaterial whether a value is assigned so long as it is understood that the corresponding terms in Riemann sums are zero.

Integrals of the form $\int_{-\infty}^{\infty} f(x)\, dx$, $\int_{-\infty}^{a} f(x)\, dx$, and $\int_{a}^{\infty} f(x)\, dx$ may be defined in exactly the language used in Section 1.2 for integrals over bounded intervals.

Again every gauge γ possesses a γ-fine division. Moreover the integral is unique. The proofs in Section S1.8 apply here.

To use the definition of the integral on an unbounded interval requires techniques similar to those which apply to $\int_{a}^{b} f(x)\, dx$ when a and b are finite but f is unbounded near a or b. Unboundedness of the interval does call for a new kind of majorant for the error made in approximating the integral over a small interval $[u, v]$ by a term $f(z)(v - u)$ of the Riemann sum. Again the objective is to insure that small individual errors do not accumulate into a large error by summation. It is not enough to have a majorant $\epsilon(v - u)$, say, on the bounded intervals. The next example illustrates what one can do.

Example 5. Show that $\int_1^\infty x^{-2} \, dx$ exists and find its value.

Solution. Since $-1/x$ is an antiderivative of x^{-2}, the usual improper integral techniques suggest that $\int_1^\infty x^{-2} \, dx = 1$. The objective of the example is to define a gauge γ on $[1, \infty]$ so that $|1 - fL(\mathfrak{D})| < \epsilon$ when $\mathfrak{D}$ is γ-fine, $f(x) = x^{-2}$ for $1 \leqslant x < \infty$, and $f(\infty)$ is given any value we like.

Since $\int_u^v x^{-2} \, dx = -1/v - (-1/u)$, a majorant is needed for $-1/v + 1/u - z^{-2}(v - u)$ when $u \leqslant z \leqslant v < \infty$. A suitable majorant is obtained by factoring out $v - u$, but not $(v - u)^2$. Straightforward calculations yield

$$\left| -\frac{1}{v} + \frac{1}{u} - \frac{1}{z^2}(v - u) \right| \leqslant \frac{v - u}{z} \left(\frac{1}{u} - \frac{1}{v} \right).$$

The factor $1/u - 1/v$ in this majorant is helpful because the sum of $1/u - 1/v$ over the bounded intervals of a division of $[1, \infty]$ does not exceed 1. Indeed, if $1 < t < \infty$ and $\mathfrak{E}$ is a division of $[1, t]$,

$$\sum_{z[u, v] \in \mathfrak{E}} \left(\frac{1}{u} - \frac{1}{v} \right) = 1 - \frac{1}{t}$$

since this is a telescoping sum.

Set $\gamma(z) = (z - \epsilon z/4, z + \epsilon z/4)$ when $1 \leqslant z < \infty$ and $\gamma(\infty) = (2/\epsilon, \infty]$. When $z \in [u, v]$ and $[u, v] \subseteq \gamma(z)$ for a finite z we have

$$\left| -\frac{1}{v} + \frac{1}{u} - \frac{1}{z^2}(v-u) \right| \leq \frac{\epsilon}{2}\left(\frac{1}{u} - \frac{1}{v} \right).$$

On the other hand, $[t, \infty] \subseteq \gamma(\infty)$ implies

$$\left| \frac{1}{t} - f(\infty)L([t, \infty]) \right| < \frac{\epsilon}{2}$$

since $L([t, \infty]) = 0$ by decree.

Let $\mathcal{D}$ be γ-fine. Since $\infty \in \gamma(z)$ only when $z = \infty$, the unbounded interval in $\mathcal{D}$, say $[t, \infty]$, has ∞ as its tag. Thus $[t, \infty] \subseteq \gamma(\infty)$.

The combination of the preceding facts shows us readily that $|1 - fL(\mathcal{D})| < \epsilon$ when $\mathcal{D}$ is γ-fine.

As an aside we may note that the bounded intervals of a γ-fine division need not all be small ones in this example. For instance, let us choose n so that $(1 + \epsilon/5)^{n-1} > 2/\epsilon$ and set $x_k = (1 + \epsilon/5)^k$ for $k = 0, 1, \ldots, n-1$ and $x_n = \infty$. Also set $z_k = x_{k-1}$ for $k = 1, 2, \ldots, n-1$ and $z_n = \infty$. It is easy to confirm that this is a γ-fine division of $[1, \infty]$. The choice of n implies that $x_{n-1} - x_{n-2} \geq 2/5$ no matter how small ϵ may be.

1.4. The fundamental theorem of calculus. In its usual formulation the fundamental theorem of calculus asserts that $\int_a^b f(x)\, dx = F(b) - F(a)$ under the assumption that f is Riemann integrable on $[a, b]$ and that $F'(x) = f(x)$ for all x in $[a, b]$. Examples 1–5 do not meet these conditions. Yet it is clear in some of them at least that an antiderivative plays a major role. By recalling the essential points of the solutions of Examples 2, 4, and 5 we can see how these examples foreshadow a significant strengthening of the fundamental theorem through the use of the generalized Riemann integral.

In Example 2 we showed that $\int_0^1 f(x)\,dx = 2$ when $f(0) = 0$ and $f(x) = 1/\sqrt{x}$ for $0 < x \leqslant 1$. Set $F(x) = 2\sqrt{x}$ for $0 \leqslant x \leqslant 1$. Then $F'(x) = f(x)$ when $0 < x \leqslant 1$ but $F'(0)$ does not exist. Nevertheless we were able to show that $\int_0^1 f(x)\,dx = F(1) - F(0)$. Let's recall how it was done. For $0 < z \leqslant 1$ we examined $2\sqrt{v} - 2\sqrt{u} - (1/\sqrt{z})$ $\cdot (v - u)$ with $u \leqslant z \leqslant v$. Of course this expression is $F(v) - F(u) - f(z)(v - u)$. We succeeded in showing that $|F(v) - F(u) - f(z)(v - u)|$ has $(v - u)^2/(z\sqrt{z})$ as majorant when $z \in [u, v]$. The calculations did not refer explicitly to the fact that $F'(z) = f(z)$. However, such a majorant is possible only when $F'(z) = f(z)$. One can see this by a limit calculation. At $z = 0$ we used continuity of F to secure an appropriate majorant of $|F(v) - F(u) - f(z)(v - u)|$.

Example 4, where $f(x) = K$ except possibly on a countably infinite set C, is also subject to interpretation as an instance of the fundamental theorem with $F(x) = Kx$. The continuity of F was the crucial property at the points where it was not assumed that $F'(x) = f(x)$. In order to use a property of F as weak as continuity it was essential that C not have too many points. Hence we assumed C to be *countably* infinite.

The novel aspect of Example 5 was that the integration was on an unbounded interval. The functions were $f(x) = x^{-2}$ for $1 \leqslant x < \infty$ and $F(x) = -x^{-1}$ for $1 \leqslant x < \infty$ with $F(\infty) = 0$. The analysis paralleled that in Example 2, resting on the equation $F'(x) = f(x)$ when $1 \leqslant x < \infty$ and the continuity of F at the exceptional point $x = \infty$. In order to offset the unboundedness of the interval it was important to have in the majorant of $|F(v) - F(u) - f(z)(v - u)|$ a factor which has bounded sums over any bounded subinterval of $[1, \infty]$, no matter how large its upper endpoint may be. For these functions $1/u - 1/v$ proved to be a natural choice.

These examples point to a fundamental theorem with the following notable features. First, the integrability of f is proved rather than being assumed. Second, $f(x)$ need not equal $F'(x)$ for every x. A countably infinite exceptional set may be allowed so long as F is required to be continuous on $[a, b]$. Third, the interval $[a, b]$ need not be bounded.

For ease in stating and discussing this kind of fundamental theorem it is helpful to have a piece of terminology which is not quite standard.

Definition.   Let I be an interval in $\overline{\mathbf{R}}$. Let $f : I \to \mathbf{R}^q$ be given. A function $F : I \to \mathbf{R}^q$ is a *primitive* of f on I provided (i) F is continuous on I and (ii) $F'(x) = f(x)$ for all x in I except possibly a finite or countably infinite set of values of x.

Note that when F is a primitive on an unbounded interval I each infinite endpoint of I which is also a point of I will certainly be among the exceptional points mentioned in (ii). This is inevitable since the notion of derivative is meaningful only at points in $\mathbf{R}$. Even though primitives will be used initially for closed intervals, later needs have been kept in mind in framing the definition so that I is not necessarily closed.

FUNDAMENTAL THEOREM.   *If $f : [a, b] \to \mathbf{R}^q$ has a primitive F on $[a, b]$, f is integrable and $\int_a^b f(x)\, dx = F(b) - F(a)$.*

A proof will be given in Section S1.9. It requires the ideas which have been used in the examples, especially Examples 4 and 5. Of course the arguments must be stripped of the particularities of the functions used in the examples.

This version of the fundamental theorem, like the weaker one which is generally known, is a powerful tool for the evaluation of integrals. Since it asserts the existence of the integral, it is also a source of information about integrability of functions. For instance, it can obviously be used to assert the integrability of polynomials on bounded intervals. The same is true of other elementary functions where explicit formulas for primitives are known. (One customarily gets these facts from a general theorem on existence of the integral of a function which is continuous on a closed and bounded interval.) Examples 2 and 5 testify to the possibility of proving the existence of integrals which are improper integrals in the Riemann sense through this fundamental theorem. A full examination of improper integrals is given in the next section.

A fundamental theorem which asserts integrability also lends itself to some uses as a substitute for the mean value theorem of differential calculus. The next exercise is a case in point.

Exercise 4. Let F and G be continuous on an interval I. Suppose $F'(x) = G'(x)$ for all x except possibly those in a countably infinite subset of I. Show that there is a constant K such that $F(x) = G(x) + K$ for all x in I.

1.5. The status of improper integrals. An integral $\int_a^b f(x)\, dx$ is customarily termed *improper* when it does not exist in the sense of the definition in Section 1.1 but can be made meaningful as a limit, e.g., $\lim_{s \to a} \int_s^b f(x)\, dx$ or $\lim_{t \to b} \int_a^t f(x)\, dx$. Of course f must be Riemann integrable on $[s, b]$ for all $s > a$ in the first instance and on all $[a, t]$ with $t < b$ in the second. Example 2 illustrates the first of these and Example 5 the second. In both examples we have seen that an integral which is improper in the Riemann sense can be an ordinary integral in the generalized Riemann sense.

The fundamental theorem of Section 1.4 allows us to dispose of a large class of examples of this kind. Specifically, when f has a primitive F on $(a, b]$ and $\lim_{s \to a} F(s)$ exists (as a *finite* limit), we set $F(a) = \lim_{s \to a} F(s)$. The result is a primitive of f on $[a, b]$. Then, by applying the fundamental theorem first on $[a, b]$, then on $[s, b]$,

$$\int_a^b f(x) \, dx = F(b) - F(a) = F(b) - \lim_{s \to a} F(s)$$

$$= \lim_{s \to a} (F(b) - F(s)) = \lim_{s \to a} \int_s^b f(x) \, dx.$$

Consequently $\int_a^b f(x) \, dx$ exists as a generalized Riemann integral and has the same value that it would be assigned as an improper Riemann integral.

Suppose that a function fails to have an integral according to the improper integral definition. For instance $\int_0^1 x^{-2} \, dx$ is not a meaningful improper integral since $\lim_{s \to 0} \int_s^1 x^{-2} \, dx = \lim_{s \to 0} (-1 + s^{-1}) = \infty$. Does the generalized Riemann integral also fail to exist under these circumstances? The fundamental theorem cannot supply the answer to this question since it contains a sufficient condition for existence of the integral but no necessary condition. The expected answer follows from the following exercise.

Exercise 5. Suppose $\int_a^b f$ exists and also $\int_s^b f$ exists for all s such that $a < s < b$. Show that $\int_a^b f = \lim_{s \to a} \int_s^b f$.

The existence of $\int_0^1 x^{-2} \, dx$ as a generalized Riemann integral is denied by Exercise 5 since $\lim_{s \to 0} \int_s^1 x^{-2} \, dx$ is not finite.

Note that Exercise 5 could just as well be stated for $\lim_{t \to b} \int_a^t f$. It should also be kept in mind that Exercise 5 contains a redundant assumption since existence of $\int_a^b f$ implies integrability of f on all closed subintervals of $[a, b]$. (This proposition will be considered in Section 2.3.)

Exercise 5 and the fundamental theorem taken together suffice for the treatment of most of the improper integrals. One gap remains. Does $\int_a^b f$ exist when $\lim_{s \to a} \int_s^b f$ exists even in the absence of a primitive of f on $(a, b]$? It does. Thus the following statement summarizes the facts.

THEOREM.   *Let* $f : [a, b] \to \mathbf{R}^q$ *have an integral on* $[s, b]$ *for all* s *such that* $a < s < b$. *Then* $\int_a^b f$ *exists if and only if* $\lim_{s \to a} \int_s^b f$ *exists. Moreover* $\int_a^b f = \lim_{s \to a} \int_s^b f$.

The proof that f is integrable on $[a, b]$ when $\lim_{s \to a} \int_s^b f$ exists has a number of technical details in it. It may be found in Section S2.8.

This theorem tells us two things about improper integrals. First, in circumstances in which there is usually a resort to the improper integral in calculus courses the generalized Riemann integral exists precisely when the improper integral exists. Moreover the two have the same value. Second, the generalized Riemann integral on intervals in $\overline{\mathbf{R}}$ has no improper extensions.

The practical effect of the fact that improper Riemann integrals are proper generalized Riemann integrals is not very great. The calculations which reveal whether such an integral is meaningful and yield its value remain about as they were.

The application of the above theorem is not limited to those cases in which f or its primitive misbehaves at just one endpoint of $[a, b]$. When there are difficulties at both endpoints or at a point strictly between a and b the usual device can be employed. For instance, consider $\int_{-1}^1 x^{-3} \, dx$. Since the difficulty lies at $x = 0$ it is helpful to examine $\int_0^1 x^{-3} \, dx$ and $\int^0 x^{-3} \, dx$. Each of these is readily shown to have no meaning. To pass to the conclusion that $\int_{-1}^{1} x^{-3}\, dx$ is also without meaning it is necessary to employ the fact that integrability on $[a, b]$ implies integrability on all subintervals of $[a, b]$.

The following exercise provides an opportunity to apply the ideas of this section and also possesses intrinsic interest. It will receive further attention later.

Exercise 6. Let $a_1, a_2, a_3, \ldots$ be given. Set $f(x) = a_k$ when $k - 1 \leqslant x < k$ for $k = 1, 2, 3, \ldots$. Show that $\int_0^\infty f(x)\, dx$ exists if and only if $\sum_{k=1}^\infty a_k$ is a convergent series and that the integral is equal to the series.

1.6. Multiple integrals. Functions defined on intervals in spaces of two or more dimensions are the next candidates for integration. Intervals, divisions, gauges, and the other terms used before can be defined so that the definition of the integral is as before.

The spaces $\mathbf{R}^p$ and $\overline{\mathbf{R}}^p$ consist of p-tuples $(x_1, x_2, \ldots, x_p)$ of elements of $\mathbf{R}$ and $\overline{\mathbf{R}}$, respectively. An element of $\overline{\mathbf{R}}^p$ is *finite* when all its components are real numbers and *infinite*, or a *point at infinity*, when at least one component is $-\infty$ or ∞.

An interval I in $\overline{\mathbf{R}}^p$ is a Cartesian product $I_1 \times I_2 \times \cdots \times I_p$ of intervals in $\overline{\mathbf{R}}$. It is open when each of $I_1, I_2, \ldots, I_p$ is open. Closed intervals and bounded intervals are defined similarly.

An interval in $\overline{\mathbf{R}}$ is nondegenerate when it has distinct endpoints. An interval I in $\overline{\mathbf{R}}^p$ is nondegenerate when all of $I_1, I_2, \ldots, I_p$ are nondegenerate intervals in $\overline{\mathbf{R}}$. Thus it is degenerate when at least one of $I_1, I_2, \ldots, I_p$ is degenerate.

Intervals I and J are *overlapping* when $I \cap J$ is nondegenerate. This is equivalent to saying that $I \cap J$ contains a nonempty open interval.

It is helpful to interpret all the discussion of $\overline{\mathbf{R}}^p$ using pictures for $p = 2$ and $p = 3$. Of course an interval in $\mathbf{R}^2$ is just a rectangle with sides parallel to the coordinate axes. In $\overline{\mathbf{R}}^2$ the closed intervals include also half-planes, strips, quarter-planes, and half-strips with edges parallel to coordinate axes. Each of these has points at infinity, of course.

It is sometimes helpful to think of $\overline{\mathbf{R}}^p$ in terms of a bounded model. The next exercise suggests a way to do this.

Exercise 7. Let $h(x) = x/(1 + |x|)$ for real x, $h(-\infty) = -1$, and $h(\infty) = 1$.

(a) Use h to visualize $\overline{\mathbf{R}}$ and its intervals within $[-1, 1]$.

(b) Set $H(x) = (h(x_1), h(x_2), \ldots, h(x_p))$ for all x in $\overline{\mathbf{R}}^p$. Show that H maps $\overline{\mathbf{R}}^p$ in a one-to-one fashion onto $[-1, 1] \times [-1, 1] \times \cdots \times [-1, 1]$ and preserves intervals.

(c) Use the preceding part to visualize $\overline{\mathbf{R}}^2$ as a square in $\mathbf{R}^2$. Mark the scale of $\overline{\mathbf{R}}^2$ as a contracting scale on the edges of the square.

A *division* of I is a finite collection of nonoverlapping nondegenerate intervals whose union is I. Figure 3 shows three divisions of a 2-dimensional interval. They increase in generality from (a) through (c).

A *tagged interval* is a pair zJ where J is a closed interval and $z \in J$. A *tagged division* of I consists of pairs zJ with $z \in J$ and the J's forming a division of I.

  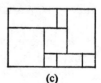

(a)   (b)   (c)

Fig. 3

A *gauge* γ is a function such that $\gamma(z)$ is an open interval containing z. A collection of tagged intervals is *γ-fine*, or *compatible with γ*, provided $J \subseteq \gamma(z)$ for each zJ in the collection.

In order to define Riemann sums we must have a *measure* of intervals corresponding to the length function L we have used in $\overline{\mathbf{R}}$. We want to define it so that it gives the usual area and volume for bounded intervals in $\mathbf{R}^2$ and $\mathbf{R}^3$. Set

$$M(I) = L(I_1)L(I_2) \cdots L(I_p)$$

when $I = I_1 \times I_2 \times \cdots \times I_p$. This measure M has the value 0 on every unbounded interval I because I_j is unbounded for some j and thus $L(I_j) = 0$. It clearly has the desired value on bounded intervals in $\mathbf{R}^2$ and $\mathbf{R}^3$.

When $\mathcal{D}$ is a tagged division of I, the Riemann sum of a function $f : I \to \mathbf{R}^q$ on $\mathcal{D}$ is

$$fM(\mathcal{D}) = \sum_{zJ \in \mathcal{D}} f(z)M(J).$$

The only change in the previous definition of the integral is the replacement of the length L by the more general measure M. The integral on an interval I is denoted

$$\int_I f, \qquad \int_I f(x)\, dx, \quad \text{or} \quad \int_I f\, dM.$$

For multidimensional spaces there are the obvious questions faced earlier in $\overline{\mathbf{R}}$. Does every gauge γ possess a γ-fine division? Is the integral unique? Yes, in both cases. A new proof is needed for the compatibility theorem. The proof of it will be deferred to Section S6.3. (See p. 168.) In the meantime we take it as proved. The uniqueness proof given in Section S1.8 is applicable to integrals on intervals in $\overline{\mathbf{R}}^p$.

In two or more dimensions it seems natural to want to integrate over sets which are not intervals. This need arises in elementary calculus in finding areas and volumes, for instance. Suppose we want to integrate f over a set E. One customary way to do so is this. Take an interval I that contains E. (It could be the whole space $\overline{\mathbf{R}}^p$.) Define a new function g which is the same as f on E and zero outside E. Define $\int_E f$ to be $\int_I g$. A caution is in order. For this definition to be fully satisfactory, it should be genuinely an extension of the notion of integration on intervals. That is, if E is itself an interval it should be true that $\int_E f$, defined using Riemann sums for divisions of E, is the same as $\int_I g$ for any interval I containing E. On page 113 we will confirm this.

The extension of f which is zero outside E may have discontinuities which f does not have as a function on E. For instance, let $f(x) = 1$ when $0 \leqslant x \leqslant 1$ and $f(x) = 0$ elsewhere. Then f is continuous when restricted to $[0, 1]$ but it has a discontinuity at 1 as a function on $[0, 2]$. Such discontinuities are usually harmless since a function need not be highly regular in order to be integrable in the generalized Riemann sense. (Recall the function which is zero on the rationals and one on the irrationals in $[0, 1]$.)

Integration over sets which are not intervals is examined at some length in Chapter 4.

The fundamental theorem given in Section 1.4 is remarkable because it asserts the integrability of f and at the same time gives a way to evaluate $\int_a^b f$. In two and more dimensions we often have to deal separately with the existence and evaluation of integrals. The most common tool for evaluation is the iterated integral. Suppose $I = [0, 1] \times [2, 4]$ and f is a well-behaved function on I. You will recall that $\int_I f$ can be expressed in two ways as an iterated integral:

$$\int_I f = \int_0^1 \left[\int_2^4 f(x, y) \, dy \right] dx = \int_2^4 \left[\int_0^1 f(x, y) \, dx \right] dy.$$

Thus the evaluation is accomplished by reducing the problem to repeated use of the fundamental theorem. What condition assures us that $\int_I f$ equals the iterated integrals? Existence of $\int_I f$ suffices. The general result to this effect is called Fubini's Theorem. It is given in Chapter 6.

Existence of iterated integrals alone does not guarantee the integrability of f. Since Fubini's Theorem asserts the equality of the iterated integrals, any function whose iterated integrals fail to be equal must be one which is not integrable. The following is a standard example.

Example 6. Calculate both iterated integrals of $f(x, y) = (x^2 - y^2)(x^2 + y^2)^{-2}$ over $[0, 1] \times [0, 1]$.

Solution. Except at $(0, 0)$ where the functions are not defined,

$$\frac{\partial}{\partial x} \left[\frac{-x}{x^2 + y^2} \right] = \frac{x^2 - y^2}{(x^2 + y^2)^2}.$$

Consequently, when $0 < y \leqslant 1$, $-x/(x^2 + y^2)$ is a primitive on $[0, 1]$ of the function $x \to f(x, y)$. Therefore,

$$\int_0^1 f(x, y) \, dx = -1/(1 + y^2)$$

when $0 < y \leqslant 1$. Whether this integral even exists for $y = 0$ is unimportant for the second integration. Since

$-\arctan y$ is a primitive for $-1/(1 + y^2)$,

$$\int_0^1 \left[\int_0^1 f(x, y)\, dx \right] dy = \int_0^1 (-1)(1 + y^2)^{-1}\, dy = -\pi/4.$$

Since $f(x, y) = -f(y, x)$, we deduce from the preceding calculation that

$$\int_0^1 \left[\int_0^1 f(x, y)\, dy \right] dx = \pi/4.$$

There are conditions which can be imposed on the iterated integral which will insure integrability. We will return to this question in Section 6.2.

1.7.   Sum of a series viewed as an integral.   In Exercise 5 we noted a way to relate convergence of a series $\sum_{k=1}^{\infty} a_k$ to existence of an integral over the interval $[0, \infty]$ in $\overline{\mathbf{R}}$. There is an alternative way to formulate series convergence as integration which is more natural in the sense that it deals entirely with sequences rather than functions on intervals of real numbers. The resulting integral is rather simple to work with. It can give guidance as to what is possible or impossible in the more complicated world of integrals over intervals in $\overline{\mathbf{R}}$.

Let $\overline{\mathbf{N}}$ be the positive integers augmented by ∞. Define intervals in $\overline{\mathbf{N}}$ by using inequalities just as in $\overline{\mathbf{R}}$. Note that apart from intervals $[m, \infty)$ every interval in $\overline{\mathbf{N}}$ is both open and closed. For instance $\{x \in \overline{\mathbf{N}} : 2 < x < 5\}$ is the same as $\{x \in \overline{\mathbf{N}} : 3 \leqslant x \leqslant 4\}$. Hence $[3, 4]$ is at the same time an open interval and a closed interval.

To define an integral on $\overline{\mathbf{N}}$ which agrees with series convergence we must define a suitable measure of intervals. The appropriate one counts the number of points in bounded intervals and assigns the value zero to unbounded intervals.

Say that two closed intervals in $\overline{\mathbf{N}}$ are nonoverlapping when they have no common points.

From these notions we can define divisions, tagged divisions, and Riemann sums in the same language as in the preceding section.

The only sort of gauge we need is one such that $\gamma(z) = \{z\}$ when $z < \infty$ and $\gamma(\infty) = [m, \infty]$ for some integer m. A γ-fine division $\mathcal{D}$ consists of one-element intervals $\{1\}, \{2\}, \dots, \{n\}$ and an unbounded interval $[n + 1, \infty]$ with $n + 1 \geqslant m$. The corresponding Riemann sum is $\sum_{k=1}^{n} a_k$ when $f(k) = a_k$ for all integers k. From these observations it is readily apparent that $\int_{\overline{\mathbf{N}}} f$ exists if and only if $\sum_{k=1}^{\infty} a_k$ is convergent and that the two of them are equal.

Taken by itself this formulation of series convergence has some interest and utility since it can offer guidance as to what is provable concerning integrals in general. It gains interest when we use this idea to define an integral on $\overline{\mathbf{N}} \times I$ where I is a closed interval in $\overline{\mathbf{R}}^p$. A function f defined on $\mathbf{N} \times I$ is equivalent to a sequence of functions f_n on I since we may let $f(n, x) = f_n(x)$. Fubini's theorem says that

$$\int_{\overline{\mathbf{N}} \times I} f = \sum_{n=1}^{\infty} \int_{I} f(n, y) \, dy = \int_{I} \left[\sum_{n=1}^{\infty} f(n, y) \right] dy$$

provided that f is integrable over $\overline{\mathbf{N}} \times I$. Consequently any condition which insures the integrability of f on $\overline{\mathbf{N}} \times I$ translates into a theorem on term-by-term integration of series.

The idea of Exercise 5 can also be used to link

integration of a series of functions on I to integration on $[0, \infty] \times I$. The choice between integration over $\overline{N} \times I$ and $[0, \infty] \times I$ is a matter of convenience.

Term-by-term integration need not be linked to Fubini's theorem. The more direct approach is initiated in Section 3.5.

S1.8. The limit based on gauges. The definition of the generalized Riemann integral is a species of limit definition. Several of the properties which make this a satisfactory limit process have been asserted in earlier sections with a promise of proofs in this section. The first such assertion was the compatibility theorem in Section 1.1. Its main thrust is that the limit process itself is always meaningful, quite apart from the question of whether any specific limit exists. Its proof is the same for intervals in $\overline{R}$ as for those in R. As noted earlier a separate proof must be given for intervals in $\overline{R}^p$ with $p \geqslant 2$. It may be found in Section S6.3.

COMPATIBILITY THEOREM.   *Let* $[a, b] \subseteq \overline{R}$ *and let* γ *be a gauge on* $[a, b]$. *There is a* γ-*fine tagged division of* $[a, b]$.

Proof. We define a subset E of $[a, b]$ as follows. Say that $x \in E$ when $a < x \leqslant b$ and there is a tagged division of $[a, x]$ which is γ-fine. There are numbers in E. To find one take x in $\gamma(a)$ with $a < x$. Tag $[a, x]$ with a. The result is a γ-fine tagged division of $[a, x]$.

Let y be the least upper bound of E. There is x in E such that $x \in \gamma(y)$ and $x < y$. By definition of E, there is a γ-fine tagged division of $[a, x]$. Adjoin to it $[x, y]$ tagged with y. The result is a γ-fine tagged division of $[a, y]$. Thus $y \in E$.

Is it possible that $y < b$? Suppose so. Let $w \in \gamma(y) \cap (y, b)$. Adjoin to any tagged division of $[a, y]$ which is γ-fine the interval $[y, w]$ tagged with y. Then $w \in E$, contrary to the assumption that y is an upper bound of E. Thus $y = b$.

As a corollary to this theorem we can conclude that there are infinitely many γ-fine tagged divisions of $[a, b]$. One of the ways to draw this conclusion is to select x in $\gamma(a)$ with $a < x < b$. Apply the theorem to $[x, b]$ to get a γ-fine division of $[x, b]$. Adjoin to it the interval $[a, x]$ tagged with a. The result is a γ-fine division of $[a, b]$. Varying x clearly produces an infinite set of distinct γ-fine divisions of $[a, b]$.

Another essential of a satisfactory limit is the uniqueness of the limit. The proof in this instance is the same whatever the dimension of the space $\overline{\mathbf{R}}^p$ in which the interval is found.

UNIQUENESS THEOREM.　*The function* $f : I \to \mathbf{R}^q$ *has at most one integral on* I.

Proof.　Let A_1 and A_2 be integrals of f on I. To show $A_1 = A_2$ it suffices to show $|A_1 - A_2| = 0$.

Let ϵ be a positive number. By definition there is a gauge γ_1 such that $|A_1 - fM(\mathcal{D}_1)| < \epsilon$ when $\mathcal{D}_1$ is a γ_1-fine division of I and a gauge γ_2 such that $|A_2 - fM(\mathcal{D}_2)| < \epsilon$ when $\mathcal{D}_2$ is a γ_2-fine division of I.

Is there a division $\mathcal{D}$ which is both γ_1-fine and γ_2-fine? Yes, it is obtained through a third gauge γ. For each z in I set $\gamma(z) = \gamma_1(z) \cap \gamma_2(z)$. Then γ is a gauge because the intersection of two open intervals containing z is an open interval containing z. Let $\mathcal{D}$ be γ-fine. (The compatibility theorem enters here.) Since $\gamma(z) \subseteq \gamma_1(z)$ for all z the division $\mathcal{D}$ is also γ_1-fine. Likewise it is γ_2-fine.

The triangle inequality gives us

$$|A_1 - A_2| \leqslant |A_1 - fM(\mathcal{D})| + |fM(\mathcal{D}) - A_2|$$

and from this it is clear that $|A_1 - A_2| < 2\epsilon$. Since ϵ is arbitrary the nonnegative number $|A_1 - A_2|$ must be zero.

Within the proof of the uniqueness theorem there is a relation and a related technique of general significance.

The relation is that one gauge γ is *stricter than* another gauge γ_1 when $\gamma(z) \subseteq \gamma_1(z)$ for all z. The useful consequence is that every γ-fine division is also γ_1-fine.

In the proof we saw how to form a gauge γ which is stricter than gauges γ_1 and γ_2. Of course this generalizes at once to gauges $\gamma_1, \gamma_2, \ldots, \gamma_n$ since we can set

$$\gamma(z) = \gamma_1(z) \cap \gamma_2(z) \cap \cdots \cap \gamma_n(z).$$

With this tool it is possible to achieve the simultaneous satisfaction of any finite set of conditions which can be satisfied severally through the choice of appropriate gauges.

S1.9. Proof of the fundamental theorem.

The essentials of the proof of the fundamental theorem are implicit in the examples given in earlier sections, particularly in Examples 4 and 5. Of course it will be necessary to rely on basic concepts rather than special algebraic expressions in giving the proof.

In order to show that $\int_a^b f$ is $F(b) - F(a)$ when F is a primitive of f we first express $F(b) - F(a) - fL(\mathcal{D})$ as a sum of terms of the form $F(v) - F(u) - f(z)L(J)$ and then estimate these terms according to whether $F'(z) = f(z)$ or whether z is in the exceptional set C on which we are sure only that F is continuous.

There is nothing but the definition of the derivative at

our disposal for the estimation of $F(v) - F(u) - f(z) \cdot (v - u)$ when $F'(z) = f(z)$. Keep in mind that we need only concern ourselves with $[u, v]$ containing z. When $u = z$ or $v = z$, the desired estimate comes easily from the definition of $F'(z)$. When $u < z < v$, i.e., when u and v straddle z, a bit more effort is required.

STRADDLE LEMMA. *Suppose $F : [a, b] \to \mathbf{R}^q$ is differentiable at z. Then there corresponds to each ϵ a δ such that*

$$|F(v) - F(u) - F'(z)(v - u)| < \epsilon|v - u|$$

whenever $z \in [u, v]$ and $[u, v] \subseteq (z - \delta, z + \delta) \cap [a, b]$.

Proof. The definition of the derivative asserts the existence of δ such that

$$\left| \frac{F(x) - F(z)}{x - z} - F'(z) \right| < \epsilon$$

when $0 < |x - z| < \delta$ and $x \in [a, b]$. Multiplication of both sides by $|x - z|$ converts this inequality into

$$|F(x) - F(z) - F'(z)(x - z)| < \epsilon|x - z|.$$

The conclusion of the lemma differs from this only in notation when $z = u$ or $z = v$. Thus we may pass on to the case $u < z < v$. Addition and subtraction of appropriate terms followed by application of the triangle inequality allows us to say that

$$\begin{aligned} |F(v) &- F(u) - F'(z)(v - u)| \\ &\leqslant |F(v) - F(z) - F'(z)(v - z)| \\ &\quad + |F(u) - F(z) - F'(z)(u - z)|. \end{aligned}$$

Each of the terms on the right may be estimated by means

of the inequality next above. Consequently

$$|F(v) - F(u) - F'(z)(v - u)| < \epsilon(|v - z| + |u - z|).$$

Since $u < z < v$, $|v - z| + |u - z| = |v - u|$. Thus the argument is complete.

A comment is in order. It is not possible to drop the requirement that $z \in [u, v]$ and simply require that u and v be close to z. Think of the behavior of the slopes of chords of a highly oscillatory graph. A specific counterexample is suggested in Exercise 10 at the end of the chapter.

The straddle lemma has a corollary which can be used to give added force to the application of the straddle lemma itself.

COROLLARY. *Suppose* $G'(z) > 0$. *There exists* δ *such that* $G'(z)(v - u) < 2(G(v) - G(u))$ *when* $z \in [u, v]$ *and* $[u, v] \subseteq (z - \delta, z + \delta)$.

Proof. We apply the straddle lemma with G in place of F. The number ϵ can be given a special value, namely $G'(z)/2$. Thus

$$|G(v) - G(u) - G'(z)(v - u)| < \tfrac{1}{2}G'(z)(v - u)$$

when $z \in [u, v]$ and $[u, v] \subseteq (z - \delta, z + \delta)$. The removal of the absolute values yields two inequalities and one of them simplifies to $G(v) - G(u) > G'(z)(v - u)/2$. This is equivalent to the desired conclusion.

This corollary is useful in securing majorants which have bounded sums over divisions of unbounded intervals. Thus it will be brought into play to accomplish the same ends we were able to reach in Example 5 through the special algebraic forms of the two functions f and F.

FUNDAMENTAL THEOREM. *If* $f : [a, b] \to \mathbf{R}^q$ *has a primitive* F *on* $[a, b]$, f *is integrable and* $\int_a^b f(x)\, dx = F(b) - F(a)$.

Proof. A notational device will be useful. Set $\Delta F(J) = F(v) - F(u)$ when $J = [u, v]$. Then

$$F(b) - F(a) = \sum_{zJ \in \mathcal{D}} \Delta F(J)$$

when $\mathcal{D}$ is a division of $[a, b]$ since the sum telescopes. Consequently

$$|F(b) - F(a) - fL(\mathcal{D})| \leqslant \sum_{zJ \in \mathcal{D}} |\Delta F(J) - f(z)L(J)|.$$

Let C be a set such that $F'(z) = f(z)$ when $z \in [a, b] - C$. Even if C is initially finite there is no harm in adding points so that C becomes infinite with $C = \{c_1, c_2, c_3, \ldots \}$. For each n there is $\gamma(c_n)$ such that $|f(c_n)L(J)| < \epsilon/2^{n+3}$ and $|\Delta F(J)| < \epsilon/2^{n+3}$ when $c_n \in J$ and $J \subseteq \gamma(c_n)$. This choice utilizes the continuity of F at c_n, of course. It implies that $|\Delta F(J) - f(c_n)L(J)| < \epsilon/2^{n+2}$ when $c_n \in J$ and $J \subseteq \gamma(c_n)$.

Since $[a, b]$ may be unbounded, a bounded auxiliary function G which is continuous on $\overline{\mathbf{R}}$ with $G'(x) > 0$ for all x in $\mathbf{R}$ will be helpful. We may use $G(x) = x/(2 + 2|x|)$ when $-\infty < x < \infty$, $G(-\infty) = -1/2$, and $G(\infty) = 1/2$.

Now assume $z \in [a, b] - C$. In applying the straddle lemma we are free to let ϵ depend on z. When we put $(\epsilon/4)G'(z)$ in place of ϵ the conclusion is

$$|F(v) - F(u) - f(z)(v - u)| < \frac{\epsilon}{4} G'(z)(v - u)$$

when $z \in [u, v]$ and $[u, v] \subseteq [a, b] \cap (z - \delta, z + \delta)$. We want to replace $G'(z)(v - u)$ by $2(G(v) - G(u))$ through application of the corollary. All that is required is to make sure δ is small enough so that the conclusions of the straddle lemma and its corollary are valid. With this δ we set $\gamma(z) = (z - \delta, z + \delta)$. Hence

$$|F(v) - F(u) - f(z)(v - u)| \leqslant \frac{\epsilon}{2}\,(G(v) - G(u))$$

when $z \in [u, v]$ and $[u, v] \subseteq [a, b] \cap \gamma(z)$.

Let $\mathfrak{D}$ be a γ-fine division of $[a, b]$. Let $\mathfrak{E}$ and $\mathfrak{F}$ be the subsets of $\mathfrak{D}$ consisting of zJ with $z \in C$ and $z \in [a, b] - C$, respectively.

When $zJ \in \mathfrak{E}$, $z = c_n$ for some n and $|\Delta F(J) - f(z) \cdot L(J)| < \epsilon/2^{n+2}$. Each c_n is the tag for at most two intervals. Consequently

$$\sum_{zJ \in \mathfrak{E}} |\Delta F(J) - f(z)L(J)| < 2 \sum_{n=1}^{\infty} \epsilon/2^{n+2} = \epsilon/2.$$

Set $\Delta G(J) = G(v) - G(u)$ when $J = [u, v]$. We can assert that

$$\sum_{zJ \in \mathfrak{F}} |\Delta F(J) - f(z)L(J)| \leqslant \sum_{zJ \in \mathfrak{F}} \frac{\epsilon}{2}\,\Delta G(J)$$

$$\leqslant \sum_{zJ \in \mathfrak{D}} \frac{\epsilon}{2}\,\Delta G(J)$$

since $\Delta G(J) > 0$ for all J. This last sum equals $(\epsilon/2)(G(b) - G(a))$ and does not exceed $\epsilon/2$ since $G(b) - G(a) \leqslant G(\infty) - G(-\infty) = 1$.

In summary, $|F(b) - F(a) - fL(\mathfrak{D})| < \epsilon/2 + \epsilon/2$ when $\mathfrak{D}$ is γ-fine. Since ϵ is arbitrary this shows that f is integrable and

$$\int_a^b f(x)\,dx = F(b) - F(a).$$

1.10. Exercises.

8. Let g be integrable on $[a, b]$. Let $f(x) = g(x)$ except possibly on a countable set C. Show that f is integrable on $[a, b]$ with the same integral as g.

9. Apply the fundamental theorem to evaluate the integral of each of the following.
   (a) $f(0) = 1$ and $f(x) = x^{-k}$, $0 < x \leqslant 1$, where $0 < k < 1$.
   (b) $f(x) = 2x$, $0 \leqslant x \leqslant 1$, and $f(x) = -1$, $1 < x \leqslant 2$.
   (c) $f(x) = \sin x$ when x is irrational and $f(x) = x$ when x is rational, $0 \leqslant x \leqslant \pi$.
   (d) $f(x) = x^{-k}$ for $x \geqslant 1$, where $k > 2$.

10. Let $f(0) = 0$ and $f(x) = x^2 \cos \pi/x$ when $x \neq 0$. Use this function and $z = 0$ to show that the straddle lemma may fail for intervals $[u, v]$ which do not straddle z.

11. Let f and g have primitives on $[a, b]$. Let k be a constant. Show that kf and $f + g$ have primitives on $[a, b]$.

12. Let $f(0) = 0$ and $f(x) = 1/\sqrt{x}$ for $0 < x \leqslant 1$. Show that f has a primitive on $[0, 1]$ but $f \cdot f$ does not.

13. Determine which of the following are meaningful.

$$\text{(a)} \int_0^\infty \sin x \, dx, \quad \text{(b)} \int_0^\infty \frac{1}{x(x+1)} \, dx, \quad \text{(c)} \int_1^\infty \frac{g(x)}{x} \, dx$$

where $g(x) = (-1)^k$ when $k - 1 \leqslant x < k$.

14. Let F be the primitive of f on $[a, b]$. Suppose there are numbers $c_0, c_1, c_2, \ldots, c_n$ such that $a = c_0 < c_1 < c_2 < \cdots < c_n = b$ and f is positive in (c_0, c_1), negative in (c_1, c_2), positive in (c_2, c_3), etc. Find a primitive for $|f|$ on $[a, b]$.

15. Modify the preceding exercise by replacing the finite sequence by an infinite sequence with $\lim_{n \to \infty} c_n = b$. Find a primitive for $|f|$ on $[a, b)$. Try to find conditions under which $|f|$ has a primitive on $[a, b]$.

16. Let f be continuous on $[a, b]$. Show that f is uniformly continuous on $[a, b]$ by using the compatibility theorem.

17. Suppose f is increasing at each point of $\mathbf{R}$; i.e., for each x there exists δ such that $x - \delta < u < x$ implies $f(u) < f(x)$ and $x < u < x + \delta$ implies $f(x) < f(u)$. Use the compatibility theorem to show f is increasing on $\mathbf{R}$.

18. Let f have a primitive on $[a, c]$ and on $[c, b]$. Show that f is integrable on $[a, b]$ and

$$\int_a^b f = \int_a^c f + \int_c^b f.$$

Replace the two intervals $[a, c]$ and $[c, b]$ by any division of $[a, b]$ and generalize the result.

19. Let f be a step function on $[a, b]$; i.e., there is a division of $[a, b]$ such that f is constant on the open intervals between the division points. Find the integral of f.

20. Let f have a primitive on $[a, b]$. Show that f is integrable on every closed subinterval of $[a, b]$.

21. (a) Let $(x_n)_{n=1}^\infty$ be a sequence in a closed interval I in $\overline{\mathbf{R}}$. Use the compatibility theorem to show this sequence has a convergent subsequence whose limit is in I.

(b) Show every sequence in $\overline{\mathbf{N}}$ has a convergent subsequence.

(c) Show every sequence in $\overline{\mathbf{N}} \times \overline{\mathbf{R}}^p$ has a convergent subsequence.

22. Define an integral for functions defined on $\mathbf{N} \times \mathbf{N}$. Compare it with existing notions of convergence of double series.

CHAPTER 2

BASIC PROPERTIES OF THE
INTEGRAL

The topics treated in this chapter are nearly all familiar from discussions of the Riemann integral. Where both propositions and proofs are familiar the treatment is concise with proofs omitted. In some instances the conclusions are standard but the hypotheses are less restrictive than the usual ones.

Section 2.1 treats four topics: linearity of the integral as a function of the integrand; component-by-component integration of vector functions; inequalities; and integration by parts. A few proofs are suggested as exercises.

In Section 2.2 the Cauchy criterion for sequences is used as a model for the Cauchy criterion for existence of integrals. The Cauchy criterion will be the major tool where existence of an integral is in question but no candidate for the value of the integral is at hand. This is the case in the next section.

Section 2.3 contains a proof of integrability of f on each closed interval J when $J \subseteq I$ and $\int_I f$ exists.

In Section 2.4 the purpose is to show that the integral is additive as a function of intervals. In one dimension this takes the form $\int_a^b f = \int_a^c f + \int_c^b f$ when $a < c < b$ and f is assumed integrable on $[a, c]$ and $[c, b]$. After the statement of the problem for intervals in $\overline{\mathbf{R}}^p$ the proof is

47

given first for one dimension. Then the difficulties which have been highlighted are dealt with in the general case as well.

Section 2.5 extends the notion of additivity from two intervals to any finite number. Since finite additivity is important for functions of intervals other than integrals, e.g., the measure function M, the proof is given in a form which makes no reference to integrals.

Section 2.6 focuses on integrals on intervals in $\overline{\mathbf{R}}$. For functions of the form $F(x) = \int_a^x f$ it deals with continuity and differentiability.

The change of variables for $\int_a^b f(x)\, dx$ occupies Section 2.7. This is a topic of great interest to users of integrals. It is not essential to the rest of the development of the theory of the integral and may be passed by on first reading. Both of the substitutions $t = \sigma(x)$ and $x = \tau(t)$ are considered. No distinction need be made between bounded and unbounded intervals since the generalized Riemann integral uses a single definition for both.

Section S2.8 completes a discussion left unfinished in Section 1.5. Proof is given that $\int_a^b f$ exists when $\lim_{s \to a} \int_s^b f$ exists. Some of the techniques used in the proof foreshadow major ideas announced in Section 3.1 and used repeatedly thereafter. The section closes with a warning that this type of limit theorem is peculiar to integrals on one-dimensional intervals.

2.1. The integral as a function of the integrand. The algebraic and order properties of the values of f induce similar properties in the integral. These are generally familiar from calculus. The proofs demand mainly the triangle inequality and the simultaneous imposition of several conditions through an appropriate gauge. It does not seem necessary to give them in full. Two exercises suggest proofs which the reader may try for practice in the use of gauges.

LINEARITY. *Let $f : I \to \mathbf{R}^q$ and $g : I \to \mathbf{R}^q$ be integrable. Let c be a real constant. Then cf and $f + g$ are integrable. Also $\int_I cf = c \int_I f$ and $\int_I (f + g) = \int_I f + \int_I g$. This extends to all finite linear combinations $\sum_{k=1}^n c_k f_k$ as well.*

Each function $f : I \to \mathbf{R}^q$ may be regarded as a q-tuple $(f_1, f_2, \ldots, f_q)$ of real-valued functions.

COMPONENT-BY-COMPONENT INTEGRATION. *The function $f : I \to \mathbf{R}^q$ is integrable if and only if its real-valued components are all integrable. Moreover $\int_I f = (\int_I f_1, \int_I f_2, \ldots, \int_I f_q)$.*

When f and g are real-valued we say that $f \leqslant g$ when $f(x) \leqslant g(x)$ for all x. Since the interval measure M is nonnegative the integral inherits inequalities from the integrand.

INEQUALITIES. *Suppose f and g are integrable on I. Then (a) $0 \leqslant \int_I f$ when $0 \leqslant f$, (b) $\int_I f \leqslant \int_I g$ when $f \leqslant g$, and (c) $|\int_I f| \leqslant \int_I g$ when $|f| \leqslant g$.*

Note that the function f may be vector-valued in (c). In Section 3.2 we will see that it is also true that $\int_I |f|$ exists under the conditions of (c). The integrability of $|f|$ is not an elementary result, in contrast to inequality (c).

When $|f| \leqslant g$ and $\int_I g = 0$, the integrability of f as well as $|f|$ is elementary.

ZERO INTEGRALS. *Suppose $|f| \leqslant g$ and $\int_I g = 0$. Then f is integrable and $\int_I f = 0$.*

The linearity of the integral and the fundamental theorem may be combined to give an integration by parts equation of considerable generality.

INTEGRATION BY PARTS.   *Let F and G be primitives of f and g respectively on* [a, b]. *Then fG is integrable if and only if Fg is integrable. Moreover*

$$\int_a^b fG = F(b)G(b) - F(a)G(a) - \int_a^b Fg.$$

Note that this integration by parts equation is valid for any combination of types of values for which the products fG, etc., are meaningful. That is, f and G may be real, one real and the other vector-valued, or both complex-valued.

The proof of integration by parts does not call for descent to the level of the definition. The preceding properties are proved by appeal to the definition.

Exercise 1.   Prove the linearity of the integral.

Exercise 2.   Prove inequality (c) above.

2.2. The Cauchy criterion. A sequence of real numbers $(a_i)_{i=1}^{\infty}$ has a limit in **R** if and only if for every positive ϵ there exists N such that $|a_i - a_j| < \epsilon$ whenever $i > N$ and $j > N$. This is the Cauchy criterion for convergence. It extends readily to sequences of elements in $\mathbf{R}^q$. The Cauchy criterion for sequences is both a pattern and a tool in developing a similar criterion for existence of integrals.

CAUCHY CRITERION.   *The function* $f : I \to \mathbf{R}^q$ *is integrable on I if and only if for every positive ϵ there exists a* *gauge* γ *such that* $|fM(\mathcal{D}) - fM(\mathcal{E})| < \epsilon$ *for all* γ-*fine divisions* $\mathcal{D}$ *and* $\mathcal{E}$ *of* I.

The proof of the necessity of the condition is very brief. Sufficiency requires creation of a sequence of gauges from which one suitable gauge can be selected.

Exercise 3. Prove the sufficiency of the Cauchy criterion.

2.3. Integrability on subintervals. *A function* f *which is integrable on an interval* I *is also integrable on each closed subinterval* J.

Since no candidate for the value of $\int_J f$ is at hand, the Cauchy criterion is an appropriate tool for proving existence of $\int_J f$. Thus we need to estimate $fM(\mathcal{D}) - fM(\mathcal{E})$ when $\mathcal{D}$ and $\mathcal{E}$ are divisions of J. Since it is Riemann sums for the interval I over which we have control, the line of attack is to adjoin the same set of tagged intervals to $\mathcal{D}$ and $\mathcal{E}$ to form divisions $\mathcal{D}'$ and $\mathcal{E}'$ of I. This has the effect of adding the same terms to $fM(\mathcal{D})$ and $fM(\mathcal{E})$; consequently $fM(\mathcal{D}) - fM(\mathcal{E}) = fM(\mathcal{D}') - fM(\mathcal{E}')$. The next three paragraphs are devoted to forming $\mathcal{D}'$ and $\mathcal{E}'$ from $\mathcal{D}$ and $\mathcal{E}$.

The set $I - J$ is not necessarily an interval but it is expressible as the union of a finite set $\mathcal{F}$ of non-overlapping intervals. This is immediate when $I \subseteq \overline{\mathbf{R}}$ and is easily seen when $I \subseteq \overline{\mathbf{R}}^p$ with $p \geqslant 2$. (The dotted lines in Fig. 1 break $I - J$ into eight intervals in a two-dimensional illustration. For any value of p at most $3^p - 1$ intervals are needed.)

Suppose a division $\mathcal{D}_K$ is formed for each interval K in $\mathcal{F}$. When all these intervals are lumped with a division $\mathcal{D}$ of J, the resulting collection $\mathcal{D}'$ is surely a division of I.

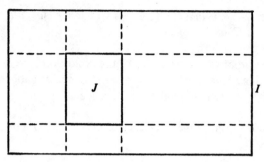

Fig. 1

Now we are ready to make appropriate choices. There is a gauge γ on I such that $|fM(\mathcal{D}') - fM(\mathcal{E}')| < \epsilon$ for any γ-fine divisions $\mathcal{D}'$ and $\mathcal{E}'$ of I. (The necessity of the Cauchy criterion for existence of $\int_I f$ assures this.) The restriction of γ to J is the gauge which guarantees $|fM(\mathcal{D}) - fM(\mathcal{E})| < \epsilon$. Indeed we need only fix a tagged division $\mathcal{D}_K$ of each K in $\mathcal{F}$. Form $\mathcal{D}'$ and $\mathcal{E}'$ as indicated above from γ-fine divisions $\mathcal{D}$ and $\mathcal{E}$ of J. Then $\mathcal{D}'$ and $\mathcal{E}'$ are γ-fine divisions of I. It follows that $|fM(\mathcal{D}) - fM(\mathcal{E})| < \epsilon$. Existence of $\int_J f$ is guaranteed by the sufficiency of the Cauchy criterion.

2.4. The additivity of integrals. *Suppose G, H and I are closed intervals such that G and H are nonoverlapping and $I = G \cup H$. Suppose f is integrable on G and on H. Then f is integrable on I and $\int_I f = \int_G f + \int_H f$.*

The most direct proof is to apply the definition of the integral. Thus $fM(\mathcal{D})$ must be shown to approximate $\int_G f + \int_H f$ well when $\mathcal{D}$ is a suitable division of I. The hypotheses enable us to use divisions of G and H to approximate the individual terms $\int_G f$ and $\int_H f$. The happiest state of affairs would be to discover that $fM(\mathcal{D}) = fM(\mathcal{E}) + fM(\mathcal{F})$ where $\mathcal{E}$ and $\mathcal{F}$ are divisions of G and H, respectively. Of course we should want $\mathcal{E}$ and $\mathcal{F}$ to inherit from $\mathcal{D}$ compatibility with appropriate gauges on G and H. This splitting of the Riemann sum can be achieved by subjecting $\mathcal{D}$ to a gauge which dictates special behavior along the common boundary of G and H. The one-dimensional case is an appropriate vehicle for getting at the essentials.

Let $G = [a, c]$ and $H = [c, b]$. In order to split a Riemann sum $fL(\mathcal{D})$ for $[a, b]$ into separate sums for $[a, c]$ and $[c, b]$, it is crucial that c be the tag for each interval in $\mathcal{D}$ which contains c. This will be the case, of course, if $\mathcal{D}$ is compatible with a gauge γ such that $c \in \gamma(z)$ only when $z = c$. In this event, $\mathcal{D}$ contains either two intervals $[u, c]$ and $[c, v]$ tagged with c or a single interval $[u, v]$ whose tag is c and whose endpoints straddle c. In the former case, $\mathcal{E}$ and $\mathcal{F}$ are formed simply by partitioning $\mathcal{D}$ in the natural way. In the latter case, the element $c[u, v]$ gives rise to an element $c[u, c]$ of $\mathcal{E}$ and an element $c[c, v]$ of $\mathcal{F}$. The remaining intervals of $\mathcal{D}$ are alloted to $\mathcal{E}$ and $\mathcal{F}$ according to whether they lie within $[a, c]$ or $[c, b]$. Clearly $\mathcal{D}$ passes along to $\mathcal{E}$ and $\mathcal{F}$ the property of being γ-fine. Splitting $\mathcal{D}$ into $\mathcal{E}$ and $\mathcal{F}$ does not quite end the story. Is $fL(\mathcal{D}) = fL(\mathcal{E}) + fL(\mathcal{F})$? This comes down to asking whether $f(c)L([u, v]) = f(c) \cdot L([u, c]) + f(c)L([c, v])$ when $[u, v]$ must be split into $[u, c]$ and $[c, v]$. Of course it is true that $L([u, v]) = L([u, c]) + L([c, v])$ when $[u, c]$ and $[c, v]$ are bounded. And this is guaranteed by insisting that $\gamma(c)$ be a bounded open interval.

In summary, these are the properties which γ must have. On G and H it must select Riemann sums which approximate $\int_G f$ and $\int_H f$ closely. When $z \notin G \cap H$ the open interval $\gamma(z)$ must contain no points of $G \cap H$. When z is finite $\gamma(z)$ must be bounded.

These properties are clearly appropriate for the one-dimensional case. They are equally successful in splitting $\mathcal{D}$ and $fM(\mathcal{D})$ in higher dimensions. Two features which arise only in $\overline{\mathbf{R}}^p$ with $p \geqslant 2$ ought to be noted. First, $\mathcal{D}$ may contain many intervals zJ which must be split because J lies neither within G nor within H. In every instance $z \in G \cap H$ because $J \subseteq \gamma(z)$ and $\gamma(z)$ contains no points of $G \cap H$ when $z \notin G \cap H$. Thus each J can be split into intervals $J \cap G$ and $J \cap H$ which may also be given the original tag z. Second, some unbounded intervals J may need splitting. When this happens z must be an infinite point because $\gamma(z)$ is bounded when z is finite. But then the pieces $J \cap G$ and $J \cap H$ are also unbounded because each contains the infinite point z. In short, splitting J always leads to one of two cases: bounded = bounded $\cup$ bounded   or   unbounded = unbounded $\cup$ unbounded,   and   never   to   unbounded = bounded $\cup$ unbounded. It is easily confirmed that $M(J) = M(J \cap G) + M(J \cap H)$ when all three intervals

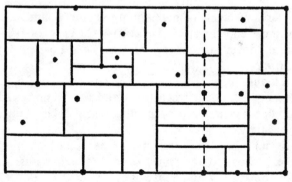

Fig. 2

are of the same type. Of course this equation is never true when unbounded = bounded ∪ unbounded because M is zero on unbounded intervals and positive on bounded ones. (Fig. 2 illustrates what may occur when I is two-dimensional. The dashed line marks the boundary between G and H. Dots indicate tags.)

The remaining details of the proof that $\int_G f + \int_H f$ is the integral of f on I are routine applications of ideas which have already become familiar.

2.5. Finite additivity of functions of intervals. Suppose I is expressed as the union of more than two nonoverlapping intervals. That is, let $\mathcal{K}$ be a division of I (without tags). Suppose $\int_K f$ exists for all K in $\mathcal{K}$. The natural extension of the additivity asserted in Section 2.4 is the claim that $\int_I f$ exists and equals $\sum_{K \in \mathcal{K}} \int_K f$.

One might think of adapting the argument used in Section 2.4 to this more general case. Indeed this is feasible. Another kind of argument is preferable, however. We want an approach that does not use the definition of the integral but builds on the two-interval additivity. Why? Because it is desirable to prove also the finite additivity of the interval measure M. That is, we want to show that $M(I) = M(\mathcal{K}) = \sum_{K \in \mathcal{K}} M(K)$ when $\mathcal{K}$ is a division of a bounded interval I.

The obvious approach is to try mathematical induction. In $\overline{\mathbf{R}}$ this succeeds with no complications. But there is a stumbling block in higher dimensions. Figure 3(a) shows what it is. Note that there is no pair of intervals in this division of I whose union is also an interval. Thus a straightforward inductive attack is blocked for completely general divisions.

There is a special type of division which lends itself to an inductive proof of finite additivity. Figure 3(b)

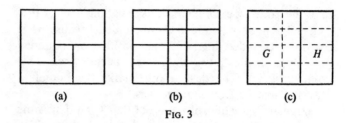

(a)   (b)   (c)

FIG. 3

illustrates it in two dimensions. For the purposes of this discussion let's refer to such divisions as regular divisions.

When $\mathcal{K}$ is a regular division of I, there are subintervals G and H such that $\{G, H\}$ is a division of I and the intervals of $\mathcal{K}$ form regular divisions $\mathcal{K}_G$ and $\mathcal{K}_H$ of G and H also. (Note Fig. 3(c) in relation to 3(b).)

Suppose finite additivity holds for regular divisions with fewer than n intervals. When $\mathcal{K}$ has n intervals, each of $\mathcal{K}_G$ and $\mathcal{K}_H$ has fewer than n intervals. Thus $M(G) = M(\mathcal{K}_G)$ and $M(H) = M(\mathcal{K}_H)$. But $M(I) = M(G) + M(H)$. Since $\mathcal{K} = \mathcal{K}_G \cup \mathcal{K}_H$ it follows that $M(I) = M(\mathcal{K})$. By the second principle of induction, finite additivity holds for all regular divisions.

Now we proceed to deal with the general case by introducing auxiliary regular divisions.

Figure 4 suggests that all the boundaries of the intervals of a given division $\mathcal{K}$ be extended. The result is a regular division $\mathcal{D}$ of I which can be grouped into regular divisions $\mathcal{D}_K$ of the intervals K of $\mathcal{K}$. Then $M(K) = M(\mathcal{D}_K)$ for all K in $\mathcal{K}$ and $M(I) = M(\mathcal{D})$ because finite additivity has been proved for regular divisions. The terms in $M(\mathcal{D})$ can be grouped so that $M(\mathcal{D}) = \sum_{K \in \mathcal{K}} M(\mathcal{D}_K)$. Thus $M(I) = \sum_{K \in \mathcal{K}} M(K) = M(\mathcal{K})$.

Note that this procedure uses not only the original intervals of $\mathcal{K}$ but also some of their subintervals.

This same argument is applicable to any function ν defined on a set $\mathcal{I}$ of intervals provided (i) $\mathcal{I}$ contains all

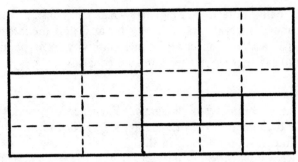

FIG. 4

closed subintervals of each of its members, (ii) $I \in \mathcal{I}$ when I possesses a division $\{G, H\}$ whose members are in $\mathcal{I}$, and (iii) $\nu(I) = \nu(G) + \nu(H)$ when $\{G, H\}$ is a division of I.

Thus, under conditions (i), (ii), and (iii), we can conclude that $\nu(I) = \sum_{K \in \mathcal{K}} \nu(K)$ when $\mathcal{K}$ is a division of I whose members are in $\mathcal{I}$. This property is *finite additivity* of ν. As already indicated, it is true of integrals and also of the interval measure M on bounded intervals. We will have occasion to use it for other interval functions, too.

It will be a harmless self-indulgence to speak of additivity where finite additivity is the property intended and the context makes this intention clear.

Finite additivity of M plays a major role in the following familiar proposition: *Let f be continuous on a bounded closed interval I. Then f is integrable, even Riemann integrable, on I.*

There will be numerous occasions for the use of the following fact.

Exercise 4. Let ν be finitely additive and nonnegative on the subintervals of I. Let $\mathcal{E}$ be a finite set of nonoverlapping subintervals of I. Show that $\nu(\mathcal{E}) \leqslant \nu(I)$.

Finally, note in passing that when f is constant on a bounded interval I, say $f(x) = K$ for all x in I, the integral of f is $KM(I)$. Indeed, the finite additivity of M permits us to say that every Riemann sum equals $KM(I)$.

2.6. Continuity of integrals. Existence of primitives.

For the remainder of this chapter we go back to the study of integrals on intervals in $\overline{\mathbf{R}}$.

Recall the *oriented integral*. It has $\int_a^b f = -\int_b^a f$ when $b < a$ and $\int_a^a f = 0$. Thus $\int_a^b f$ is meaningful whatever the relationship of a and b, provided f is integrable on some interval containing a and b. By checking all possible cases one sees that

$$\int_a^b f = \int_a^c f + \int_c^b f$$

whenever a, b, and c belong to an interval on which f is integrable.

Now suppose f is integrable on every closed subinterval of an interval H in $\overline{\mathbf{R}}$. (H need not be a closed interval.) Fix a in H and set $F(x) = \int_a^x f$ for all x in H. We are interested in the continuity and differentiability of F.

The function F is continuous on H. Since continuity of F at c is equivalent to $\lim_{x \to c} \int_a^x f = \int_a^c f$, the proof reduces to a couple of applications of Exercise 5 of Section 1.5.

A full treatment of the differentiability properties of F requires deep investigations. There is an easy partial result, however. *If f is continuous at c, then $F'(c)$ exists and $F'(c) = f(c)$.* The proof which is usually given for this statement in calculus texts is valid for the generalized Riemann integral, too. It need not be reproduced here.

These facts about continuity and differentiability make possible an assertion about the existence of a primitive for a given function f.

Suppose f is integrable on [a, b] and continuous except possibly at a countable set in [a, b]. Then f has a primitive on [a, b], namely, the function F given by $F(x) = \int_a^x f$. In particular, if f is continuous on [a, b], it is necessarily integrable and has a primitive. (It is Riemann integrable, in fact.)

Knowing a primitive exists and being able to express it in a form other than $\int_a^x f$ are very different matters, of course.

2.7. Change of variables in integrals on intervals in $\overline{\mathbf{R}}$. The change of variables technique for evaluating integrals is applied in two ways. These are illustrated in the following examples.

Example 1. Evaluate $\int_0^1 (x^2 + 1)^n x\, dx$ when $n \neq -1$.

Solution. Set $t = x^2 + 1$. Then $dt = 2x\, dx$ and

$$\int_0^1 (x^2 + 1)^n x\, dx = \int_1^2 \frac{1}{2} t^n\, dt = \frac{1}{2} \left. \frac{t^{n+1}}{n+1} \right|_1^2 = \frac{2^{n+1} - 1}{2(n+1)}.$$

Example 2. Evaluate $\int_0^1 (x^2 + 1)^{-3/2}\, dx$.

Solution. Set $x = \tan t$. Then $dx = \sec^2 t\, dt$ and

$$\int_0^1 (x^2 + 1)^{-3/2}\, dx = \int_0^{\pi/4} \frac{\sec^2 t\, dt}{(\tan^2 t + 1)^{3/2}}$$

$$= \int_0^{\pi/4} \cos t\, dt = 1/\sqrt{2}.$$

Clearly, there is a significant difference between Example 1 and Example 2. In Example 1 it is easy to spot an expression $\sigma(x)$ such that the integrand is of the form $g(\sigma(x))\sigma'(x)$. The substitution $t = \sigma(x)$ converts $\int_a^b f(x)\,dx$ into $\int_{\sigma(a)}^{\sigma(b)} g(t)\,dt$. The fundamental theorem and the chain rule for differentiation of composite functions can be used to find conditions under which this type of substitution is valid. We will look into this shortly.

In Example 2 the choice of the substitution results from noting that the integrand has a form which is simplified by a certain type of substitution. The substitution is the reverse of the type used in Example 1. That is, we replace x by a function of t rather than replacing a function of x by t. There are two ways to justify this procedure. One is to interpret it as an instance of the procedure used in Example 1 with starting integral and target integral interchanged. The other is to attack it directly by going down to the level of Riemann sums. These lead to the imposition of quite different restrictions. We will look into both.

Now let's start on the development of conditions under which the substitution $t = \sigma(x)$ may be used.

The chain rule, $(G \circ \sigma)'(x) = G'(\sigma(x))\sigma'(x)$, is valid when $G \circ \sigma$ is defined on an open interval containing x, $\sigma'(x)$ exists and $G'(t)$ exists for $t = \sigma(x)$.

When $f(x) = g(\sigma(x))\sigma'(x)$, it is clear from the chain rule that we should seek G such that $G'(t) = g(t)$ and then consider $G \circ \sigma$ as a possible primitive of f. If $G'(t) = g(t)$ for all t and $\sigma'(x)$ exists for all x in $[a, b]$ all is well. If there are exceptional values, we must proceed more cautiously, since the exceptional set must be countable for the fundamental theorem to be applicable.

Suppose $G \circ \sigma$ is a primitive of $(g \circ \sigma)\sigma'$ on $[a, b]$. Then the fundamental theorem yields

$$\int_a^b f(x)\,dx = \int_a^b g(\sigma(x))\sigma'(x)\,dx = G(\sigma(b)) - G(\sigma(a)).$$

Suppose G is a primitive of g on an interval containing $\sigma(a)$ and $\sigma(b)$. Then

$$\int_{\sigma(a)}^{\sigma(b)} g(t)\, dt = G(\sigma(b)) - G(\sigma(a)).$$

Consequently $\int_a^b f(x)\, dx = \int_{\sigma(a)}^{\sigma(b)} g(t)\, dt$ under these circumstances. Note that we may use the oriented integral here and avoid any restriction as to which of $\sigma(a)$ and $\sigma(b)$ is the larger. Indeed they may even be equal.

These observations support a proposition which may be stated in the following detailed form.

Let $f : [a, b] \to \mathbf{R}^q$ be given. Suppose there are $\sigma : [a, b] \to H$ and $g : H \to \mathbf{R}^q$ such that $f(x) = g(\sigma(x))\sigma'(x)$ except possibly on a countable subset of $[a, b]$. Suppose moreover that g has a primitive G on the interval H in $\bar{\mathbf{R}}$ such that $G \circ \sigma$ is also a primitive of $(g \circ \sigma)\sigma'$ on $[a, b]$. Then $\int_a^b f$ exists and

$$\int_a^b f = \int_a^b (g \circ \sigma)\sigma' = \int_{\sigma(a)}^{\sigma(b)} g.$$

Note that the existence of $\int_a^b f$ is asserted in this proposition. Of course in Example 1 there was the obvious ground of continuity for existence of the integral. Example 3 shows that substitution may indeed be a convenient way to confirm the integrability of a given function.

Example 3. Show that $\int_0^\infty x^{-2} e^{-1/x}\, dx$ exists and find its value.

Solution. Let $t = x^{-1}$. Then $dt = -x^{-2}\, dx$ and

$$\int_0^\infty x^{-2} e^{-1/x}\, dx = \int_\infty^0 (-e^{-t})\, dt = e^{-t} \Big|_\infty^0 = 1.$$

In the framework of the proposition above we have used $\sigma(x) = x^{-1}$ when $0 < x < \infty$, $\sigma(0) = \infty$, and $\sigma(\infty) = 0$. Thus H is $[0, \infty]$ and $G(t) = e^{-t}$ for $0 \leqslant t < \infty$ while $G(\infty) = 0$. Consequently $(G \circ \sigma)' = (g \circ \sigma)\sigma'$ on $(0, \infty)$ and G is continuous on $[0, \infty]$. Evidently $G \circ \sigma$ is a primitive on $[0, \infty]$. So is G itself. Thus this substitution is valid.

One can, of course, check substitutions individually, as in Example 3, to see whether $G \circ \sigma$ is a primitive of $(g \circ \sigma)\sigma'$. However, there are some general conclusions which cover all the cases one commonly meets.

We are assuming that G is a primitive of g on H. Thus G is continuous on H and $G' = g$ on $H - K$ where K is a countable subset of H. What conditions on σ are conducive to $G \circ \sigma$ being a primitive? So that $G \circ \sigma$ will be continuous it is proper to demand that σ be continuous. Note, however, that we may safely broaden the usual sense of continuity by allowing $\sigma(c)$ to be infinite for some points c in $[a, b]$ provided $\sigma(c) = \lim_{x \to c}\sigma(x)$. This permits the use of unbounded intervals H as in Example 3. The more delicate question is how to make the exceptional set for $(G \circ \sigma)' = (g \circ \sigma)\sigma'$ be countable.

We must put some assumption on existence of σ'. It is natural to try to allow a countable exceptional set here. Thus we suppose that σ' exists on $[a, b] - C$ where C is countable. Set $E = \{x \in [a, b] : x \notin C$ and $\sigma(x) \in K\}$. Then $(G \circ \sigma)' = (g \circ \sigma)\sigma'$ except possibly on $C \cup E$. Since C is countable, the exceptional set $C \cup E$ is also countable when E is countable. The extreme case of an empty set E is achieved when K is empty, i.e., when $G' = g$ everywhere on H. This surely occurs when H is bounded and g is continuous on H. The set E is countable when σ is one-to-one. However, this is needlessly restrictive. It is enough that σ be countable-to-one. This means that $\{x \in [a, b] : \sigma(x) = t\}$ is countable for each t.

The preceding two paragraphs may be summarized as follows.

Let $\sigma : [a, b] \to H$ be continuous on $[a, b]$ and differentiable on $[a, b] - C$ for some countable set C. (σ may have infinite values.) Let G be a primitive of g on H. Each of the following suffices for $G \circ \sigma$ to be a primitive of $(g \circ \sigma)\sigma'$ on $[a, b]$:

   (i) $G' = g$ *throughout* H,
   (ii) σ *is countable-to-one.*

The properties of the derivative can be used to check to see whether σ is countable-to-one. The following exercise gives a criterion which is easy to apply in individual instances.

Exercise 5.   Suppose $\sigma : [a, b] \to \overline{\mathbf{R}}$ is differentiable on $[a, b] - C$ and that σ' is nowhere zero on $[a, b] - C$ for some countable set C. Show that σ is countable-to-one.

Now we are ready to consider problems typified by Example 2. That is, $\int_a^b f(x)\, dx$ is given and it is proposed to set $x = \tau(t)$ and consider $\int_c^d f(\tau(t))\tau'(t)\, dt$ with $a = \tau(c)$ and $b = \tau(d)$.

As a first attack on this problem, let's interpret it as a problem of the preceding type with all roles reversed. That is, if we set $g = (f \circ \tau)\tau'$, it is as if $\int_c^d g$ were given and we propose to convert it into $\int_a^b f$. A primitive of f must play a key role in this view. But it is usually the inability to find a primitive of f which leads to the use of substitution. However, a primitive of f may be known to exist, on the strength of Section 2.6, say, without its form being known.

In fact, Section 2.6 and the prior discussions of this section, including Exercise 5, support the following assertion.

Suppose $\int_a^b f$ exists and f is continuous on $[a, b] - C$ with C countable. Suppose $\tau : [c, d] \to [a, b]$ is continuous on $[c, d]$ and, with countably many exceptions, $\tau'(t) \neq 0$. Then $\int_{\tau(c)}^{\tau(d)} f = \int_c^d (f \circ \tau)\tau'$.

Now we shift to a second attack on the problem of justifying the substitution $x = \tau(t)$ in $\int_a^b f(x)\,dx$. Instead of focusing on the fundamental theorem we move down to the basic level of the definition. What properties of τ enable us to relate Riemann sums for f to Riemann sums for $(f \circ \tau)\tau'$? When τ is one-to-one this is easily done. We have two cases according to whether τ is increasing or decreasing. To avoid some notational nuisances, let's speak only of the case where τ is increasing for the moment. By using the techniques we used in proving the fundamental theorem, we can define a gauge γ on $[c, d]$ so that $|fL(\mathfrak{D}) - gL(\mathfrak{E})| < \epsilon$ when $\mathfrak{E}$ is any γ-fine division of $[c, d]$ and $\mathfrak{D}$ is its counterpart in $[a, b]$. The continuity of τ also allows us to use a given gauge γ_1 on $[a, b]$ to restrict γ further so that $\mathfrak{D}$ is γ_1-fine when $\mathfrak{E}$ is γ-fine. The upshot of these facts is that existence of $\int_a^b f$ implies integrability of g on $[c, d]$ with $\int_a^b f = \int_c^d g$.

The notational details attendant upon the case of decreasing τ are easily disposed of. We are able to claim the following.

Suppose $\int_a^b f$ exists. Suppose $\tau : [c, d] \to [a, b]$ is continuous and one-to-one on $[c, d]$ and $\tau'(t)$ exists with countably many exceptions. Then $\int_{\tau(c)}^{\tau(d)} f = \int_c^d (f \circ \tau)\tau'$.

Compare this statement with the preceding proposition.

Clearly, the more f is restricted the less τ need be restricted and vice versa.

Note that f and $(f \circ \tau)\tau'$ do not play symmetrical roles. Integrability of f implies integrability of $(f \circ \tau)\tau'$ but not necessarily the reverse. A little more restriction on τ' will allow the implication to run both ways. Clearly, τ has an inverse function σ which is continuous and one-to-one. When $\tau'(t) \neq 0$ with countably many exceptions the derivative of σ exists with countably many exceptions. Thus σ and $(f \circ \tau)\tau'$ can be put in place of τ and f in the assertion above. Since f results from the substitution of σ and σ' into $(f \circ \tau)\tau'$, we conclude that integrability of $(f \circ \tau)\tau'$ implies integrability of f. Thus *existence of $\int_a^b f$ is equivalent to existence of $\int_c^d (f \circ \tau)\tau'$ when $\tau'(t) \neq 0$ with countably many exceptions*.

The requirement that τ be one-to-one can be relaxed to the extent of requiring that τ be one-to-one on each subinterval of a division of $[c, d]$. A proposition along these lines is offered as Exercise 18 at the end of this chapter.

S2.8. Limits of integrals over expanding intervals.
The major purpose of this section is to show that $\int_a^b f$ exists when $\lim_{s \to a} \int_s^b f$ exists. The implications of this statement were noted in Section 1.5 but the proof was deferred. Two preliminary items lead into the main result.

The first item is to note how well a gauge designed to select Riemann sums for $\int_a^b f$ performs at the like task for a subinterval. An argument reminiscent of the proof of the existence of the integral on subintervals is appropriate.

Exercise 6. Suppose $\int_a^b f$ exists and γ is a gauge such that $|\int_a^b f - fL(\mathcal{D})| < \epsilon$ when $\mathcal{D}$ is any γ-fine division of $[a, b]$. Let

$[c, d] \subseteq [a, b]$. Show that $\left| \int_c^d f - fL(\mathcal{E}) \right| < 2\epsilon$ when $\mathcal{E}$ is any γ-fine division of $[c, d]$.

The next step is to construct a gauge γ on $(a, b]$ out of gauges on subintervals of $(a, b]$ to achieve uniform approximation of $\int_s^b f$ by Riemann sums. *Assume $\int_s^b f$ exists for all s satisfying $a < s < b$. There exists a gauge γ on $(a, b]$ such that $\left| \int_s^b f - fL(\mathcal{E}) \right| < \epsilon$ for any s in (a, b) and any γ-fine division $\mathcal{E}$ of $[s, b]$.*

The construction rests on defining γ to achieve progressively closer approximations on a sequence of closed intervals which fill $(a, b]$. Begin by selecting a decreasing sequence of numbers c_n with $c_0 = b$ and $\lim_{n \to \infty} c_n = a$. Fix a gauge γ_1 on $[c_1, c_0]$ so that $\left| \int_{c_1}^{c_0} f - fL(\mathcal{D}) \right| < \epsilon/2^2$ when $\mathcal{D}$ is a γ_1-fine division of $[c_1, c_0]$. For $n \geqslant 2$, fix a gauge γ_n on $[c_n, c_{n-2}]$ so that $\left| \int_{c_n}^{c_{n-2}} f - fL(\mathcal{D}) \right| < \epsilon/2^{n+1}$ when $\mathcal{D}$ is a γ_n-fine division of $[c_n, c_{n-2}]$. (The reason for using $[c_n, c_{n-2}]$ instead of $[c_n, c_{n-1}]$ will appear shortly.)

When $z \in (c_1, c_0]$ define $\gamma(z)$ so that $\gamma(z) \subseteq \gamma_1(z)$ and $\gamma(z) \subseteq (c_1, \infty]$. When $z \in (c_n, c_{n-1}]$ with $n > 1$ define $\gamma(z)$ so that $\gamma(z) \subseteq \gamma_n(z)$ and $\gamma(z) \subseteq (c_n, c_{n-2})$. Thus γ is defined on $(a, b]$.

Let $\mathcal{E}$ be a γ-fine division of $[s, b]$ with $a < s < b$. For each n let $\mathcal{E}_n$ be the subset of $\mathcal{E}$ whose tags are in $(c_n, c_{n-1}]$. Only a finite number of the sets $\mathcal{E}_n$ are nonempty and no two of them have common elements. Let J_n be the union of the intervals belonging to $\mathcal{E}_n$. Then J_n is an interval. Moreover, the defining properties of γ imply that $\mathcal{E}_n$ is γ_n-fine, $J_1 \subseteq (c_1, c_0]$, and $J_n \subseteq (c_n, c_{n-2})$ when $n > 1$. Thus Exercise 6 is applicable and yields the conclusion $\left| \int_{J_n} f - fL(\mathcal{E}_n) \right| < 2\epsilon/2^{n+1}$. Since the integral is additive and $fL(\mathcal{E}) = \sum_n fL(\mathcal{E}_n)$, we conclude that $\left| \int_s^b f - fL(\mathcal{E}) \right| < \sum_{n=1}^{\infty} \epsilon/2^n$.

The preceding construction required no limit of the integral on $[s, b]$. The completion of γ on $[a, b]$, i.e., the definition of $\gamma(a)$, uses precisely that assumption.

For convenience let $A = \lim_{s \to a} \int_s^b f$. Choose $\gamma(a)$ so that $|f(a)L([a, s])| < \epsilon$ and $|A - \int_s^b f| < \epsilon$ when $s \in \gamma(a)$. A straightforward inequality yields $|A - fL(\mathcal{D})| < 3\epsilon$ when $\mathcal{D}$ is a γ-fine division of $[a, b]$. Thus we have completed what we set out to do, namely prove the following.

Suppose $\int_s^b f$ exists for all s in (a, b) and $\lim_{s \to a} \int_s^b f$ exists. Then f is integrable on $[a, b]$.

The next exercise points out two sets of circumstances in which this proposition can be used.

Exercise 7. Suppose $\int_s^b f$ exists for all s in (a, b).
   (a) Assume $f \geqslant 0$ also. Show that $\int_a^b f$ exists if and only if $\int_s^b f$ is bounded as s ranges over (a, b).
   (b) Suppose $|f| \leqslant g$ and $\int_a^b g$ exists. Show that $\int_a^b f$ exists also.

Note that the two parts of Exercise 7 are analogues of criteria for convergence of series. Part (a) corresponds to boundedness of partial sums of series with nonnegative terms. Part (b) is a comparison test. As it stands it is an imperfect analogue to the comparison test for absolute convergence since it does not claim existence of $\int_a^b |f|$. This blemish is removable, as we will see in Section 3.2.

Finally, a warning against excessive hopes for generalization of the limit theorem to multiple integrals. Consider the following example. Let $f(x, y) = 1$ when $0 \leqslant x \leqslant \infty$ and $0 \leqslant y \leqslant 1$. Let $f(x, y) = -1$ when $0 \leqslant x \leqslant \infty$ and $1 < y \leqslant 2$. Let $J = [0, t] \times [0, 2]$. It is not difficult to see that $\int_J f = t - t = 0$. Thus $\lim_J \int_J f = 0$ as J expands with increasing t to fill $[0, \infty) \times [0, 2]$. But it is impossible that

f be integrable on $[0, \infty] \times [0, 2]$. If so f would also be integrable on $[0, \infty] \times [0, 1]$ where it has the value 1. The latter is clearly impossible.

There are ways to generalize this limit theorem to multiple integrals. They are considered in Section 4.5. (See p. 121.)

2.9. Exercises.

8. Show that the integration by parts formula holds for the dot product in $\mathbf{R}^q$ and the cross product in $\mathbf{R}^3$. Generalize.

9. Suppose $\int_I |f| = 0$. Show that $\int_J f = 0$ for every interval J satisfying $J \subseteq I$.

10. Suppose $\int_I |f - g| = 0$. Show that f is integrable on I if and only if g is integrable on I and that $\int_J f = \int_J g$ whenever $J \subseteq I$.

11. Let P be a hyperplane given by an equation $x_i = $ constant in $\overline{\mathbf{R}}^p$. Suppose f is zero except possibly on P. Show that $\int_I f = 0$ for any interval I.

12. Let I be a closed interval and let H be its interior. Suppose $\int_I f$ exists and f vanishes outside H. Show that $\int_J f = \int_I f$ when $I \subseteq J$.

13. Suppose $\int_I f$ exists. Let $g(x) = f(x)$ when $x \in I$ and $g(x) = 0$ otherwise. Show that $\int_J g = \int_I f$ when $I \subseteq J$. Suggestion: use linearity of the integral and the preceding two exercises.

14. Use linearity of the integral and the preceding exercise to give an alternative proof of additivity: $\int_I f = \int_G f + \int_H f$ when $\{G, H\}$ is a division of I and f is integrable on each of G and H.

15. Show that $\int_0^\infty e^{-x^2}\, dx$ and $\int_0^\infty e^{-x}\sin\sqrt{x}\ dx$ exist.

16. Suppose f has a primitive on $[a, b]$ and G has a continuous first derivative everywhere on $[a, b]$. Show that $\int_a^b fG$ exists.

17. Let $\int_I |f| = 0$. Show that $\{x \in I : f(x) = 0\}$ includes all points at which f is continuous.

18. Suppose τ maps $[c, d]$ continuously onto $[a, b]$. Suppose $\tau'(t) \neq 0$ for all t in $[c, d] - F$ where F is finite. Let $f : [a, b] \to \mathbf{R}^q$ be given. Show that f is integrable on $[a, b]$ if and only if $(f \circ \tau)\tau'$ is integrable on $[c, d]$. Show, moreover, that $\int_{\tau(r)}^{\tau(s)} f = \int_r^s (f \circ \tau)\tau'$ for any $[r, s]$ contained in $[c, d]$.

19. Show that $\int_1^\infty f(x)\,dx$ exists if and only if $\int_0^1 (1/x^2) \cdot f(1/x)\,dx$ exists and that these integrals are equal.

20. Use the substitution $x = kt$ to argue that $\int_0^1 (1/x)\,dx$ is not a meaningful integral.

21. Show that $\int_0^1 (1/\sqrt{1 - x^2})\,dx$ exists in three ways: (i) by applying Exercise 7, (ii) by substitution, and (iii) by using the fundamental theorem and the derivative of $\sqrt{1 - x^2}/x$.

CHAPTER 3

ABSOLUTE INTEGRABILITY AND
CONVERGENCE THEOREMS

Some of the most useful tools of integration theory will be developed in this chapter. They center on two important operations on functions. One is the formation of the absolute value. The other is the limit of a sequence of functions. The behavior of the generalized Riemann integral with respect to these operations exhibits the strength of this integral definition most vividly.

When f is integrable, it is important to be able to tell whether $|f|$ is also integrable. A simple criterion is stated in Section 3.2. This criterion has implications for the calculation of the length of curves. These are also explored in Section 3.2. The proof of the criterion for integrability of $|f|$ is deferred to Section S3.8, since the argument contains a substantial number of technical details.

In the discussion of the integrability of $|f|$ we make use of $\sum_J |\int_J f|$ where the sum is taken over the intervals J in a division of I. The link between this sum and a Riemann sum for $|f|$ is given by a technical tool of fundamental importance called Henstock's lemma. Henstock's lemma is presented in Section 3.1, but its proof is deferred to Section S3.7. It should be added that Henstock's lemma plays a role in the proofs of nearly all the deep results about the generalized Riemann integral.

Section 3.3 utilizes the results of the preceding section and prepares the way for some later sections. It is concerned with conditions for integrability of the function which equals the larger of $f(x)$ and $g(x)$ and the function which equals the smaller of $f(x)$ and $g(x)$ where f and g are given real-valued integrable functions.

Now we turn to the second topic of the title of the chapter.

Quite often a function f is defined as, or expressed as, the limit of a sequence or the sum of a series. Say that we have $f(x) = \lim_{n \to \infty} f_n(x)$ for all x in an interval I. If the functions f_n are all integrable on I it is important to ask whether f is also integrable on I and whether $\int_I f = \lim_{n \to \infty} \int_I f_n$. Since $f = \lim_{n \to \infty} f_n$ we can also write this as

$$\int_I \left(\lim_{n \to \infty} f_n \right) = \lim_{n \to \infty} \int_I f_n.$$

The form of this last equation suggests the customary statement that the limit and the integral have been interchanged. When $f(x) = \sum_{n=1}^{\infty} f_n(x)$ for all x in I, the corresponding equation is $\int_I f = \sum_{n=1}^{\infty} \int_I f_n$, that is,

$$\int_I \left(\sum_{n=1}^{\infty} f_n \right) = \sum_{n=1}^{\infty} \int_I f_n.$$

Here one says that the integration and summation have been interchanged or that the series has been integrated term-by-term.

Whether integration can be interchanged with the limit of a sequence or the sum of a series depends upon the nature of the convergence of the sequence or series as well as on the generality of the integral being used. The

Riemann integral can be interchanged with the limit of a uniformly convergent sequence. The generalized Riemann integral can also be interchanged with the limit of a monotone sequence and of a sequence whose terms have a common integrable dominant.

The theorem on integration of uniformly convergent sequences is stated and proved in Section 3.4. The opportunity is also grasped to give a useful result connecting uniform convergence and differentiation.

Section 3.5 contains the statement of the monotone convergence theorem for sequences and its counterpart for series of nonnegative terms. An example is offered to illustrate a typical use of the theorem. The proof of the theorem is deferred to Section S3.10.

Section 3.6 deals with dominated convergence in a manner similar to Section 3.5. Again the major proof is deferred to Section S3.10.

The proofs for monotone and dominated convergence have landed in a single section because they are both based on a common principle. That principle is actually implicit in the simpler proof of integration of uniformly convergent sequences. An integral is a type of limit. We can write appropriately $\int_I f = \lim_{\mathcal{D}} f M(\mathcal{D})$ if we keep in mind that the limit process here is the one defined in Section 1.2. Now $\int_I (\lim_{n\to\infty} f_n)$ can be reformulated as $\lim_{\mathcal{D}} (\lim_{n\to\infty} f_n M(\mathcal{D}))$ and $\lim_{n\to\infty} \int_I f_n$ has the alternative form $\lim_{n\to\infty} (\lim_{\mathcal{D}} f_n M(\mathcal{D}))$. Thus interchange of integration with the sequence limit is the same as this limit interchange:

$$\lim_{\mathcal{D}} \left(\lim_{n\to\infty} f_n M(\mathcal{D}) \right) = \lim_{n\to\infty} \left(\lim_{\mathcal{D}} f_n M(\mathcal{D}) \right).$$

The all-important principle here is that equality of iterated limits occurs when one of the inner limits is uniform with respect to the other variable.

In Section 3.4 uniform convergence of f_n to f is used to show that $\lim_{n\to\infty} f_n M(\mathcal{D})$ is uniform with respect to $\mathcal{D}$. When the convergence of f_n to f is monotone or dominated, the roles are reversed and we show that $\lim_{\mathcal{D}} f_n M(\mathcal{D})$ is uniform with respect to n. This is the basis of the argument in Section S3.10.

Section S3.9 contains a general formulation of the limit concept and a proof of the theorem on interchange of iterated limits when one of the inner limits is uniform.

3.1. Henstock's lemma. A function f which is integrable on I is integrable on every subinterval of I. Thus it is meaningful and natural to ask how well $f(z)M(J)$ approximates $\int_J f$ when J is a subinterval of I containing z. More generally, given any subset $\mathcal{E}$ of a tagged division of I, what can be said about $\sum_{zJ\in\mathcal{E}}[f(z) \cdot M(J) - \int_J f]$ and about $\sum_{zJ\in\mathcal{E}}|f(z)M(J) - \int_J f|$? The answer is somewhat surprising.

HENSTOCK'S LEMMA.   *Let* $f : I \to \mathbf{R}^q$ *be integrable on* I. *Let* γ *be a gauge on* I *such that* $|fM(\mathcal{D}) - \int_I f| < \epsilon$ *when* $\mathcal{D}$ *is any* γ-*fine division of* I. *Let* $\mathcal{E}$ *be a subset of a* γ-*fine division of* I. *Then* $|\sum_{zJ\in\mathcal{E}}[fM(zJ) - \int_J f]| \leqslant \epsilon$ *and* $\sum_{zJ\in\mathcal{E}}|fM(zJ) - \int_J f| \leqslant 2q\epsilon$.

The proof will be deferred to Section S3.7.

Roughly speaking Henstock's lemma asserts that a gauge γ selects Riemann sums as well on subintervals of I as it does on the whole interval I. This enables us to break a division $\mathcal{D}$ into subsets according to some property associated with the tags without losing the close approximation of the sum of integrals by the terms of the Riemann sum. This aspect of Henstock's lemma is at the heart of the proof of the monotone and dominated convergence theorems which is given below in Section S3.10.

A second aspect of Henstock's lemma plays the dominant role in formulating and proving a criterion for integrability of $|f|$. The last conclusion of Henstock's lemma allows us to claim that $\sum_{zJ \in \mathfrak{D}} |f(z)| M(J)$ is close to $\sum_{zJ \in \mathfrak{D}} |\int_J f|$. This is a consequence of the inequality $\|A| - |B\| \leqslant |A - B|$ which holds for any elements A and B of $\mathbf{R}^q$. These ideas are pursued further in the next section and Section S3.8.

Henstock's lemma can be found in the proofs of most of the deeper assertions about the generalized Riemann integral. The next example illustrates a simple application of Henstock's lemma. The result is an interesting one which is not trivial to prove by other means.

Example 1. Suppose $\int_a^c f = 0$ for all c in $(a, b]$. Show that $\int_a^b |f| = 0$.

Solution. Choose γ so that $|fL(\mathfrak{D}) - \int_a^b f| < \epsilon$ when $\mathfrak{D}$ is γ-fine. Suppose $a < c < d \leqslant b$. Then

$$\int_c^d f = \int_a^d f - \int_a^c f = 0.$$

Consequently $|f(z)L(J) - \int_J f| = |f(z)|L(J)$ for every zJ in $\mathfrak{D}$. From Henstock's lemma, $|f|L(\mathfrak{D}) \leqslant 2q\epsilon$. This shows that $\int_a^b |f| = 0$.

A comment about subsets of divisions of I may prove helpful. When $\mathfrak{E}$ is a nonempty subset of a division $\mathfrak{D}$, it is evident that $\mathfrak{E}$ consists of a finite number of nonoverlapping intervals. Of course each has a tag attached. Conversely, *any finite set of nonoverlapping subintervals of I, each equipped with a tag, is a subset of a*

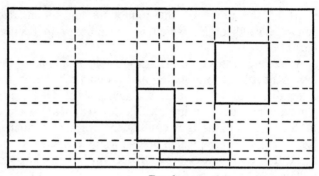

Fig. 1

tagged division of I. This is an extension of the idea presented in Section 2.3 for incorporating a single interval into a division of I. Figure 1 suggests what needs to be done first. That is, by extending the boundaries of the intervals in $\mathcal{E}$ cut the part of I not covered by the intervals of $\mathcal{E}$ into a collection $\mathcal{K}$ of nonoverlapping intervals which also do not overlap the intervals of $\mathcal{E}$. Assign a tag to each member of $\mathcal{K}$. Then $\mathcal{K} \cup \mathcal{E}$ is a tagged division of I. Of course $\mathcal{K} \cup \mathcal{E}$ may be useless for certain purposes. For instance, if $\mathcal{E}$ is γ-fine for some given γ the division $\mathcal{K} \cup \mathcal{E}$ may fail to be γ-fine. This is easily remedied. Instead of assigning a tag to each K in $\mathcal{K}$ form a γ-fine division $\mathcal{D}_K$ of each interval K in $\mathcal{K}$. Amalgamate $\mathcal{E}$ with all the divisions $\mathcal{D}_K$. The result is a γ-fine division of I which contains $\mathcal{E}$ as a subset. Of course there are many ways to form $\mathcal{D}_K$. Sometimes it is helpful to impose further conditions in forming the $\mathcal{D}_K$. This is, in fact, the key to the proof of Henstock's lemma itself.

Often a more concise way of referring to a finite set of nonoverlapping subintervals is very helpful. The term *partial division* will be used. The presence of tags can be indicated by context or by explicit reference to a *tagged partial division*.

3.2.   Integrability of the absolute value of an integrable function.   When f is integrable on I, it is certainly not automatic that $\int_I |f|$ exists, too. This is evident from the fact that series convergence is equivalent to existence of a certain kind of integral. Of course we know from examples like $\sum_{k=1}^\infty (-1)^k k^{-1}$ that convergence of $\sum_{k=1}^\infty a_k$ need not entail convergence of $\sum_{k=1}^\infty |a_k|$. In line with the terminology for series, one says that f is *absolutely integrable* on I when $\int_I f$ and $\int_I |f|$ exist.

In observations above concerning Henstock's lemma we noted that $\sum_{zJ \in \mathfrak{D}} |\int_J f|$ is closely approximated by $|f| M(\mathfrak{D})$. Thus these sums of absolute values of integrals must have a close connection with the existence and value of $\int_I |f|$. It is to such sums that we now turn our attention.

For convenience set $\nu(J) = \int_J f$ and $|\nu|(J) = |\nu(J)|$. In keeping with previous notations we let $\nu(\mathfrak{S}) = \sum_{zJ \in \mathfrak{S}} \nu(J)$ and $|\nu|(\mathfrak{S}) = \sum_{zJ \in \mathfrak{S}} |\nu|(J)$ when $\mathfrak{S}$ is a finite set of intervals.

A division $\mathfrak{S}$ is a *refinement* of a division $\mathfrak{D}$ when each interval in $\mathfrak{S}$ is a subset of an interval of $\mathfrak{D}$. Another way to put the same idea is that $\mathfrak{S} = \bigcup_{J \in \mathfrak{D}} \mathfrak{S}_J$ where each $\mathfrak{S}_J$ is a division of J. The finite additivity of ν and the triangle inequality imply at once that $|\nu(J)| \leqslant \sum_K |\nu(K)|$ where the sum runs over all members of $\mathfrak{S}_J$. Summation over all J in $\mathfrak{D}$ yields $|\nu|(\mathfrak{D}) \leqslant |\nu|(\mathfrak{S})$ when $\mathfrak{S}$ is a refinement of $\mathfrak{D}$. In short, $|\nu|$ is increasing with respect to refinement.

Subjecting a tagged division to the control of a gauge tends to make its intervals small. Passage from a division to its refinements also results in smaller intervals. The former is appropriate to approximating $\int_I |f|$ by its Riemann sums. The latter causes $|\nu|(\mathfrak{D})$ to increase, hence to approximate its least upper bound. Since both lead us to small intervals, there is reason to conjecture a least upper-bound criterion for existence of $\int_I |f|$. The result which can be obtained is the following.

INTEGRATION OF ABSOLUTE VALUES. *Suppose $f : I \to \mathbf{R}^q$ is integrable on I. Set $\nu(J) = \int_J f$ when $J \subseteq I$. Then $|f|$ is integrable on I if and only if $\mathrm{lub}_{\mathfrak{D}} |\nu|(\mathfrak{D})$ is finite. Moreover $\int_I |f| = \mathrm{lub}_{\mathfrak{D}} |\nu|(\mathfrak{D})$.*

The proof will be deferred to Section S3.8. It is direct in its basic ideas but rather technical in its details.

A comparison test often makes it easy to show that an integrable function is also absolutely integrable.

COMPARISON TEST FOR ABSOLUTE INTEGRABILITY. *Suppose f and g are integrable on I and $|f| \leqslant g$. Then $|f|$ is integrable and $|\int_I f| \leqslant \int_I |f| \leqslant \int_I g$.*

The proof brings together several ideas. When $J \subseteq I$, we can assert $|\int_J f| \leqslant \int_J g$ without knowing whether $|f|$ is integrable, as noted in Section 2.1. The finite additivity of the integral of g implies that $\sum_{J \in \mathfrak{D}} |\int_J f| \leqslant \int_I g$ for any division $\mathfrak{D}$ of I. Thus $\int_I |f|$ exists according to the criterion above. The final inequalities are direct applications of the facts stated in Section 2.1.

The comparison test implies that each real-valued component of an absolutely integrable vector-valued function is also absolutely integrable. The converse also follows by comparison test.

Sums and differences of absolutely integrable functions are also absolutely integrable by the comparison test.

A further refinement of the comparison test will be given in Section 4.5. (See p. 120.) The objective there will be to replace the assumption of integrability of f on I by integrability on each of a sequence of subsets of I. The resulting comparison test is an exact analogue of the comparison test for convergence of series. Indeed it is a generalization of the test for series convergence.

The final form of the comparison test, in terms of measurable functions, is given on page 136.

The remainder of this section deals with absolute integrability of functions on intervals in $\overline{\mathbf{R}}$.

Suppose f has a primitive F on $[a, b]$. The integrability of $|f|$ is reflected in a property of F which we will now identify.

By the fundamental theorem $\int_u^v f = F(v) - F(u)$ when $[u, v] \subseteq [a, b]$. For convenience set $\Delta F(J) = F(v) - F(u)$ when $J = [u, v]$. Existence of $\int_a^b |f|$ is equivalent to $\text{lub}_{\mathcal{D}} |\Delta F|(\mathcal{D})$ being finite according to the proposition above. The least upper bound of $|\Delta F|(\mathcal{D})$ is a familiar quantity. We recall the appropriate definition.

Definition. A function $F : [a, b] \to \mathbf{R}^q$ is a function of *bounded variation* when $\text{lub}_{\mathcal{D}} |\Delta F|(\mathcal{D})$ is finite. ($\mathcal{D}$ runs over all divisions of $[a, b]$.) The number $\text{lub}_{\mathcal{D}} |\Delta F|(\mathcal{D})$ is called the *total variation* of F on $[a, b]$.

In this language *the function f is absolutely integrable provided its primitive F is a function of bounded variation.* The next example shows that a primitive need not be a function of bounded variation.

Example 2. Construct a function whose primitive has among its values the partial sums of the alternating series $1 - 1/2 + 1/2 - 1/3 + 1/3 - \cdots$ in such fashion that the primitive is not a function of bounded variation.

Solution. We will construct a function f and its primitive F on $[a, b]$. (Assume a is finite.) Select a sequence $c_0, c_1, c_2, \ldots$ so that $a = c_0$, $b = \lim_{n \to \infty} c_n$ and $c_n < c_{n+1}$ for all n. The given series has partial sums s_n such that $s_0 = 0$, $s_{2k-1} = 1$, and $s_{2k} = 1 - 1/(k+1)$ for $k \geqslant 1$. Let $F(c_n) = s_n$ for $n = 0, 1, 2, \ldots$ and let F be linear on each interval $[c_n, c_{n+1}]$. Then F is the primitive on $[a, b]$ of the function f which is constant on each interval $[c_n, c_{n+1})$ and equal to the derivative of F on (c_n, c_{n+1}). Since $\lim_{n \to \infty} s_n = 1$, it is evident that $\lim_{x \to b} F(x) = 1$ also. Set $F(b) = 1$ to complete the definition of the primitive of f on $[a, b]$.

Let $\mathcal{D}_n$ be the division of $[a, b]$ formed by the points $c_0, c_1, c_2, \ldots, c_n, 1$. Let $n = 2m + 1$. Then

$$|\Delta F|(\mathcal{D}_n) = 1 + \frac{1}{2} + \frac{1}{2} + \frac{1}{3} + \frac{1}{3} + \cdots$$

$$+ \frac{1}{m+1} + \frac{1}{m+1}.$$

Consequently $|\Delta F|(\mathcal{D}_1) = 1$, $|\Delta F|(\mathcal{D}_3) = 2$, $|\Delta F|(\mathcal{D}_7) > 3$, $|\Delta F|(\mathcal{D}_{15}) > 4$ and, in general, $|\Delta F|(\mathcal{D}_n) \geqslant i$ when $n = 2^i - 1$. Clearly, F is not a function of bounded variation on $[a, b]$.

The relation between the integrability of $|f|$ and the properties of the primitive F is applicable to the discussion of the length of curves in $\mathbf{R}^q$.

When $q \geqslant 2$, a continuous function $g : [a, b] \to \mathbf{R}^q$ can be interpreted as a parametric representation of a curve in $\mathbf{R}^q$. The length of the curve is defined by inscribing polygonal curves and finding the least upper bound of the length of such inscribed polygons. The inscribing of the polygon is achieved by choosing division points

$$a = x_0 < x_1 < \cdots < x_n = b$$

and joining consecutive points $g(x_0), g(x_1), \ldots, g(x_n)$ by line segments. The length of the segment from $g(x_{k-1})$ to

$g(x_k)$ is just $|g(x_k) - g(x_{k-1})|$. Thus the length of the inscribed polygon is $|\Delta g|(\mathcal{D})$ when $\mathcal{D}$ is the division $\{[x_0, x_1], \ldots, [x_{n-1}, x_n]\}$. Consequently, the length of the curve and the total variation of g are identical and the curve has finite length if and only if g is a function of bounded variation.

Now suppose g has a derivative except possibly at a countable subset of $[a, b]$. The length of the curve given by g is thus $\int_a^b |g'|$. With $g = (g_1, g_2, \ldots, g_q)$ we have $g' = (g_1', g_2', \ldots, g_q')$ and thus

$$\text{length} = \int_a^b \left[(g_1')^2 + (g_2')^2 + \cdots + (g_q')^2 \right]^{1/2}.$$

This is the familiar arc length formula when $q = 2$ and $q = 3$. In summary, *when g is continuous on $[a, b]$ and g' exists except possibly on a countable set, the curve given by g has a finite length if and only if $\int_a^b |g'|$ exists. Moreover $\int_a^b |g'|$ is its length.*

Example 3. Let $g(t) = (t, F(t))$ where F is the primitive constructed in Example 2. Then g gives a parametric representation of the graph of F. Show that the graph of F does not have finite length.

Solution. Since $|\Delta F(J)| < |\Delta g(J)|$ for every interval J it is immediate that $|\Delta F|(\mathcal{D}) < |\Delta g|(\mathcal{D})$ for every division $\mathcal{D}$ of $[a, b]$. Since F does not have bounded variation the same is true of g. Consequently the curve defined by g, i.e., the graph of f, does not have finite length.

Of course $|g'|$, or $[1 + (F')^2]^{1/2}$, is not integrable on $[a, b]$.

3.3. Lattice operations on integrable functions. The formation of the least upper bound, or supremum, and the greatest lower bound, or infimum, of a set of two real-valued functions is another operation whose relation to integrability is important.

Definition. Let f and g be real functions. On the common part of their domains set

$$(f \vee g)(x) = \max\{f(x), g(x)\},$$
$$(f \wedge g)(x) = \min\{f(x), g(x)\}.$$

Note that $f \vee g$ may be neither f nor g, since $f(x)$ may be larger than $g(x)$ for some x while the reverse holds for other x. There is a very useful way to express $f \wedge g$ and $f \vee g$ in terms of $|f - g|$, namely,

$$f \vee g = \tfrac{1}{2}[f + g + |f - g|],$$
$$f \wedge g = \tfrac{1}{2}[f + g - |f - g|].$$

Check these by confirming the corresponding equations in which f and g have been replaced by real numbers u and v. Since linear combinations of integrable functions are integrable, $f \vee g$ and $f \wedge g$ are integrable when f, g, and $|f - g|$ are integrable.

Example 4. Let $f^+ = f \vee 0$ and $f^- = (-f) \vee 0$ where 0 denotes the zero function. Then $f^+ = (f + |f|)/2$ and $f^- = (-f + |f|)/2$. A little algebra gives $f = f^+ - f^-$, while $|f| = f^+ + f^-$. From these four equations it is clear that integrability of f and $|f|$ is equivalent to integrability of f^+ and f^-.

Since there are integrable functions which are not absolutely integrable, the supremum of integrable functions is not always integrable. The same is true of infimum since $f \wedge 0 = -f^-$.

The functions f^+ and f^- are called the *positive part* and *negative part* of f, respectively. Notice that each is nonnegative, however.

Let f, g, h be integrable on I. Suppose f ⩽ h and g ⩽ h, or h ⩽ f and h ⩽ g. Then f ∨ g and f ∧ g are integrable.

Suppose first that h is a common majorant, i.e., $f \leqslant h$ and $g \leqslant h$. Then $f \vee g \leqslant h$ also. Consequently

$$|f - g| = 2(f \vee g) - f - g \leqslant 2h - f - g.$$

Since $f - g$ is integrable and $|f - g|$ has an integrable dominant, it follows from the comparison test that $|f - g|$ is integrable. Thus $f \vee g$ and $f \wedge g$ are integrable.

The second case is handled similarly by showing that $|f - g| \leqslant f + g - 2h$.

The definition of the infimum and the supremum can be extended to finite sets of functions in a straightforward way. Then an induction argument extends the criterion for integrability as follows.

Let $f_1, f_2, \ldots, f_n$ and h be integrable. Suppose $f_k \leqslant h$ for all k or $h \leqslant f_k$ for all k. Then $\bigwedge_{k=1}^{n} f_k$ and $\bigvee_{k=1}^{n} f_k$ are integrable.

Note that $h = |f_1| + \cdots + |f_n|$ is a suitable dominant when each f_k is absolutely integrable. Thus the lattice operations yield integrable functions when applied to absolutely integrable functions.

3.4. Uniformly convergent sequences of functions. Recall that $\lim_{n \to \infty} f_n(x) = f(x)$ *uniformly on a set E* provided there exists n_ϵ such that $|f_n(x) - f(x)| < \epsilon$ for all x in E and all $n > n_\epsilon$. Uniform convergence is the condition usually given in discussions of the Riemann integral to justify $\lim_{n \to \infty} \int_I f_n = \int_I (\lim_{n \to \infty} f_n)$. Uniform convergence remains as the basis of the simplest proof of the interchange of the sequence limit with the generalized Riemann integral.

UNIFORM CONVERGENCE THEOREM.   *Suppose*

$$\lim_{n\to\infty} f_n(x) = f(x)$$

uniformly on the bounded interval I. Let each f_n be integrable on I. Then f is integrable on I and $\lim_{n\to\infty}\int_I f_n = \int_I f$.

The first step in the proof is to show that the sequence of integrals is a Cauchy sequence. Choose n_ϵ so that $|f_n(x) - f(x)| < \epsilon$ for all x in I and all $n > n_\epsilon$. Then $|f_m(x) - f_n(x)| < 2\epsilon$ and $|\int_I f_m - \int_I f_n| \leqslant 2\epsilon M(I)$ when m and n exceed n_ϵ.

Note also that $|f_n M(\mathcal{D}) - fM(\mathcal{D})| = |(f_n - f)M(\mathcal{D})| \leqslant \epsilon M(\mathcal{D})$ for any division $\mathcal{D}$ when $n > n_\epsilon$. Since I is bounded $M(\mathcal{D}) = M(I)$. Thus $f_n M(\mathcal{D})$ converges to $fM(\mathcal{D})$ uniformly in $\mathcal{D}$.

Now let $A = \lim_{n\to\infty}\int_I f_n$ and select n so that $n > n_\epsilon$ and $|\int_I f_n - A| < \epsilon$. Choose a gauge γ_n so that $|\int_I f_n - f_n M(\mathcal{D})| < \epsilon$ when $\mathcal{D}$ is γ_n-fine. For a γ_n-fine $\mathcal{D}$,

$$|A - fM(\mathcal{D})| \leqslant \left| A - \int_I f_n \right| + \left| \int_I f_n - f_n M(\mathcal{D}) \right|$$
$$+ |f_n M(\mathcal{D}) - fM(\mathcal{D})|$$
$$< \epsilon + \epsilon + \epsilon M(I).$$

Consequently $A = \int_I f$.

Note that this argument shows that f is Riemann integrable when the members of the sequence are Riemann integrable.

The restriction to bounded intervals is an essential limitation.

Exercise 1. Give an example which shows that uniform convergence of f_n to f does not imply integrability of f when I is unbounded.

By using the uniform convergence theorem and an additional argument of the same type we can prove a theorem on term-by-term differentiation.

Suppose F_n is a primitive of f_n on the bounded interval $[a, b]$. Let $(f_n)_{n=1}^{\infty}$ converge uniformly to f on $[a, b]$. Suppose that $\lim_{n \to \infty} F_n(c)$ exists for some c in $[a, b]$. Then $(F_n)_{n=1}^{\infty}$ converges uniformly on $[a, b]$ to a function F which is a primitive of f. Moreover, $F'(x) = f(x)$ whenever $F_n'(x) = f_n(x)$ for all n.

Exercise 2. Prove the preceding proposition.

The satisfaction of the condition on the values of F_n at a point c is easily arranged. The addition of a constant to a primitive produces another primitive. If primitives F_n are given so that $(F_n(c))_{n=1}^{\infty}$ is not convergent, the addition of suitable constants to the functions F_n will cause the values at c to converge for the modified primitives. (For instance, $-F_n(c)$ can be added to F_n.) Consequently, f has a primitive on the bounded interval $[a, b]$ if f is the uniform limit of functions f_n having primitives on $[a, b]$.

This assertion can be used to give a new proof of the existence of a primitive for a continuous function. The next exercise calls attention to another class of functions having primitives.

Exercise 3. Show that f has a primitive on the bounded interval $[a, b]$ when f is increasing on $[a, b]$ or decreasing on $[a, b]$.

In Section S3.9 the essential features of the proof given above for the uniform convergence theorem will be placed in a more general framework. The result will be an iterated limits theorem of wide applicability.

3.5. The monotone convergence theorem. A sequence of functions $(f_n)_{n=1}^{\infty}$ is *monotone* when it is increasing or decreasing. It is *increasing* if $f_n(x) \leqslant f_{n+1}(x)$ for all n and x. Reversal of the inequality yields a *decreasing* sequence. A monotone sequence has, for each x, only two options for limit behavior. That is, $\lim_{n \to \infty} f_n(x)$ exists as a finite limit or as an infinite limit. The version of the monotone convergence theorem which follows assumes a finite limit exists everywhere. In Section 4.2 we will note why this assumption can be removed.

MONOTONE CONVERGENCE THEOREM. *Let $(f_n)_{n=1}^{\infty}$ be a monotone sequence with $\lim_{n \to \infty} f_n(x) = f(x)$ for all x in I. Then $\int_I f$ exists if and only if $\lim_{n \to \infty} \int_I f_n$ is finite. Moreover $\int_I f = \lim_{n \to \infty} \int_I f_n$.*

The interchange of integration with the sequential limit in this theorem will be shown to be a consequence of uniform convergence too. In these circumstances the convergence of $f_n M(\mathcal{D})$ to $\int_I f_n$ is uniform with respect to n. By contrast, in the preceding section we used the uniform convergence of $f_n M(\mathcal{D})$ to $f M(\mathcal{D})$. In both instances a special kind of convergence of $f_n(x)$ to $f(x)$ gives rise to the uniformity of a limit of $f_n M(\mathcal{D})$.

Henstock's lemma plays a major role in showing that the convergence of $f_n M(\mathcal{D})$ to $\int_I f_n$ is uniform with respect to n. The detailed argument is offered in Section S3.10.

The following example exhibits a typical application of monotone convergence.

Example 5. Suppose f and g are nonnegative functions on the interval I. Suppose f is integrable on I and g is integrable on each bounded subinterval of I. Show that $f \wedge g$ is integrable on I.

Solution. Fix expanding bounded intervals I_n so that $\bigcup_{n=1}^{\infty} I_n = I \cap \mathbf{R}^p$. Let g_n equal g on I_n and zero elsewhere. Set $h_n = f \wedge g_n$. Then h_n is integrable on I since f and g_n are integrable functions with zero as integrable minorant. Furthermore, h_n increases to $f \wedge g$ on I provided we take the harmless precaution of setting $f(x) = 0$ when x is an infinite point of I. Since $\int_I h_n \leqslant \int_I f$, the monotone convergence theorem asserts the integrability of $f \wedge g$ on I.

Note that monotone convergence allows for the interchange of integration with the sequence limit without limiting the interval I in any way.

Now we shall translate the monotone convergence theorem for sequences into appropriate series language. A series $\sum_{n=1}^{\infty} f_n$ converges when its sequence of partial sums converges. Since the partial sum s_n is $\sum_{k=1}^{n} f_k$ and $s_n - s_{n-1} = f_n$, the partial sums form an increasing sequence when $f_n \geqslant 0$ for $n \geqslant 2$. Similarly, the partial sums decrease when the terms are nonpositive. Consequently the monotone convergence theorem can be recast as follows.

INTEGRATION OF SERIES OF POSITIVE TERMS. *Let f_n be integrable on I for $n = 1, 2, 3, \ldots$ and let $f_n \geqslant 0$. Suppose $\sum_{n=1}^{\infty} f_n(x)$ is finite for all x in I. Then $\sum_{n=1}^{\infty} f_n$ is integrable on I if and only if $\sum_{n=1}^{\infty} \int_I f_n$ is finite. Moreover $\sum_{n=1}^{\infty} \int_I f_n = \int_I (\sum_{n=1}^{\infty} f_n)$.*

The monotone convergence theorem and its counterpart for series give further evidence of the difference between the Riemann and the generalized Riemann integrals. An earlier example can be used to show that the Riemann integral does not have such theorems.

Example 6. Let $(r_n)_{n=1}^{\infty}$ be a sequential arrangement of the rationals in $[0, 1]$, with $r_m \neq r_n$ when $m \neq n$. Let $f_n(x) = 1$ when $x = r_n$ and zero otherwise. Then $\int_0^1 f_n$ exists in the Riemann sense and has the value zero. Thus $\sum_{n=1}^{\infty} \int_0^1 f_n$ is convergent but $\sum_{n=1}^{\infty} f_n$ is not Riemann integrable since it is zero on the irrationals and one on the rationals in $[0, 1]$.

3.6. The dominated convergence theorem.

It is important to note that not even the generalized Riemann integral allows free interchange of integral and sequential limit.

Example 7. Let $f_n(x) = n$ when $x \in (0, 1/n)$ and $f_n(x) = 0$ otherwise. Then $\lim_{n \to \infty} f_n(x) = 0$ for all x in $[0, 1]$. However, $\int_0^1 f_n = 1$ for all n. Consequently, $\lim_{n \to \infty} \int_0^1 f_n$ and $\int_0^1 (\lim_{n \to \infty} f_n)$ both exist, but they fail to be equal.

The dominated convergence theorem rules out behavior such as that exhibited in Example 7 by a condition which is simple to state and easy to check in specific instances.

LEBESGUE'S DOMINATED CONVERGENCE THEOREM. *Suppose f_n and h are integrable on I and $|f_n| \leq h$ for all n. Also suppose that $\lim_{n \to \infty} f_n(x) = f(x)$ for all x in I. Then f is integrable on I and $\lim_{n \to \infty} \int_I f_n = \int_I f$.*

The common integrable majorant for all the members of the sequence ensures that $f_n M(\mathcal{D})$ converges to $\int_I f_n$ uniformly with respect to n. The proof of uniformity is largely the same as that needed for the monotone convergence theorem. In Section S3.10 the common part of the proof is given first without distinguishing the way in which f_n converges to f. Then the proof is completed for monotone convergence. Lastly the monotone convergence theorem is used to complete the proof of the dominated convergence theorem.

Note that each function $|f_n|$ is also integrable under the conditions of the dominated convergence theorem by the comparison test. Consequently $|f|$ is also integrable since the dominated convergence theorem also applies to the sequence $(|f_n|)_{n=1}^{\infty}$.

A slightly more general version of the dominated convergence theorem can be given. In place of $|f_n| \leqslant h$ one can assume $|f_n - f_m| \leqslant h$ for all m and n. The argument given in Section S3.10 is more naturally connected with the latter hypothesis. This hypothesis does not imply absolute integrability of all the functions. One need only add a function which is not absolutely integrable to all members of a sequence satisfying $|f_n| \leqslant h$ to establish this point.

The functions f_n may be vector valued in both versions of the dominated convergence theorem.

By applying the dominated convergence theorem to the sequence of partial sums we get the following series formulation.

Suppose f_n and h are integrable and $|\sum_{k=1}^{n} f_k| \leqslant h$ for all n. Suppose also that $\sum_{n=1}^{\infty} f_n(x)$ converges for all x. Then $\sum_{n=1}^{\infty} \int_I f_n = \int_I (\sum_{n=1}^{\infty} f_n)$.

In Section 4.2 we will obtain a theorem on series in which convergence of $\sum_{n=1}^{\infty} \int_I |f_n|$ is the only restriction on the series. Both the monotone convergence theorem and the dominated convergence theorem will play a role in the proof.

The dominated convergence theorem will also prove useful in other parts of the chapters which follow.

When the functions f_n have a common bound, say B, on

a bounded interval I the constant function with value B can be used as the dominant h in the theorem above for sequences. This proposition is sometimes called the *bounded convergence theorem*.

S3.7. Proof of Henstock's lemma. In Henstock's lemma a gauge γ is given for which $|fM(\mathcal{D}) - \int_I f| < \epsilon$ for each γ-fine division $\mathcal{D}$ of I. The first conclusion to be reached is that $|\sum_{zJ \in \mathcal{E}}[fM(zJ) - \int_J f]| \leqslant \epsilon$ when $\mathcal{E}$ is a subset of a γ-fine division $\mathcal{D}$ of I.

There are many divisions other than $\mathcal{D}$ which contain $\mathcal{E}$ and are also γ-fine. One way to generate another is to form a division $\mathcal{D}_K$ of each interval K in $\mathcal{D} - \mathcal{E}$. Each $\mathcal{D}_K$ should be chosen to give a very close approximation of $\int_K f$ by $fM(\mathcal{D}_K)$.

For convenience let $\mathcal{F}$ be the set of intervals in $\mathcal{D} - \mathcal{E}$. Let n be the number of intervals in $\mathcal{F}$. Let ϵ' be an arbitrary positive number. Choose $\mathcal{D}_K$ so that $|fM(\mathcal{D}_K) - \int_K f| < \epsilon'/n$ and so that $\mathcal{D}_K$ is γ-fine. Let $\mathcal{D}'$ be the union of $\mathcal{E}$ and all $\mathcal{D}_K$'s. Then $\mathcal{D}'$ is a γ-fine division of I.

To make the notation manageable set $\Phi(zJ) = f(z) \cdot M(J) - \int_J f$ and extend Φ to sets of tagged intervals by summation. Then

$$\Phi(\mathcal{E}) = \Phi(\mathcal{D}') - \sum_{K \in \mathcal{F}} \Phi(\mathcal{D}_K).$$

The terms in $\Phi(\mathcal{D}')$ and $\Phi(\mathcal{D}_K)$ can be grouped as a difference of sums rather than a sum of differences. Since the integral is additive $\Phi(\mathcal{D}') = fM(\mathcal{D}') - \int_I f$ and $\Phi(\mathcal{D}_K) = fM(\mathcal{D}_K) - \int_K f$. Consequently $|\Phi(\mathcal{D}')| < \epsilon$ and $|\Phi(\mathcal{D}_K)| < \epsilon'/n$. Thus $|\Phi(\mathcal{E})| < \epsilon + \epsilon'$. Since ϵ' is an arbitrary positive number unrelated to $\mathcal{E}$, $|\Phi(\mathcal{E})| \leqslant \epsilon$.

The second conclusion of Henstock's lemma is that $\sum_{zJ \in \mathcal{E}}|fM(zJ) - \int_J f| \leqslant 2q\epsilon$. It reads $|\Phi|(\mathcal{E}) \leqslant 2q\epsilon$ when

$|\Phi|(zJ) = |f(z)M(J) - \int_J f|$ and $|\Phi|$ is extended to sets of intervals by summation.

Suppose first that f is real-valued. Let $\mathcal{E}^+$ be the set of zJ in $\mathcal{E}$ for which $\Phi(zJ) \geq 0$. Then $\Phi(\mathcal{E}^+) = |\Phi|(\mathcal{E}^+)$. Let $\mathcal{E}^-$ be the rest of $\mathcal{E}$. Then $-\Phi(\mathcal{E}^-) = |\Phi|(\mathcal{E}^-)$. The inequality which has already been proved applies to $\mathcal{E}^+$ and $\mathcal{E}^-$ since they are also subsets of $\mathcal{D}$. Thus

$$|\Phi|(\mathcal{E}) = |\Phi|(\mathcal{E}^+) + |\Phi|(\mathcal{E}^-)$$
$$= |\Phi(\mathcal{E}^+)| + |\Phi(\mathcal{E}^-)| \leq 2\epsilon.$$

Let $f = (f_1, f_2, \ldots, f_q)$. Each f_i also satisfies the hypothesis of Henstock's lemma, since the absolute value of a component of a vector does not exceed the norm of the vector. For emphasis: only one γ is being used for f and all f_i. Thus $|\Phi_i|(\mathcal{E}) \leq 2\epsilon$ for $i = 1, 2, \ldots, q$. But $|\Phi(zJ)| \leq \sum_{i=1}^q |\Phi_i(zJ)|$ also. Summation over zJ in $\mathcal{E}$ followed by reversal of the order of summation on the right gives $|\Phi|(\mathcal{E}) \leq \sum_{i=1}^q |\Phi_i|(\mathcal{E})$. Replacement of each term by 2ϵ gives the final conclusion.

S3.8. Proof of the criterion for integrability of $|f|$.
The notations of Section 3.2 will be used.

The necessary condition for absolute integrability is rather easy to prove. For any J, $|\int_J f| \leq \int_J |f|$ when f and $|f|$ are integrable. Thus $|\nu|(\mathcal{D}) \leq \int_I |f|$ since the integral of $|f|$ is additive over the division $\mathcal{D}$. Clearly $\mathrm{lub}_{\mathcal{D}} |\nu|(\mathcal{D})$ is finite.

The challenging part is to prove that $A = \int_I |f|$ when $A = \mathrm{lub}_{\mathcal{D}} |\nu|(\mathcal{D})$ and $A < \infty$. The line of attack is to show directly from the definition of the integral that A is $\int_I |f|$. The difficulty lies in the need to bring $|\nu|(\mathcal{D})$ close to A by refinement while bringing $|\nu|(\mathcal{D})$ close to $|f|M(\mathcal{D})$ by subjecting $\mathcal{D}$ to a gauge.

For a given ϵ there is a division $\mathcal{K}$ of I such that $A - \epsilon < |\nu|(\mathcal{K}) \leqslant A$. The value of $|\nu|$ on any refinement of $\mathcal{K}$ lies between $|\nu|(\mathcal{K})$ and A. When γ is chosen so that $|\int_I f - f M(\mathcal{D})| < \epsilon$ for γ-fine $\mathcal{D}$, the last conclusion of Henstock's lemma implies that $||f|M(\mathcal{D}) - |\nu|(\mathcal{D})| \leqslant 2q\epsilon$. The rest would be easy if it were possible to define γ so that any γ-fine division is also a refinement of $\mathcal{K}$. A bit less is sufficient and feasible. Through added restrictions on γ each γ-fine $\mathcal{D}$ possesses an associate $\mathcal{D}'$ with three valuable properties: $\mathcal{D}'$ is a refinement of $\mathcal{K}$, $\mathcal{D}'$ is γ-fine, and $|f|M(\mathcal{D}) = |f|M(\mathcal{D}')$. Then

$$|A - |f|M(\mathcal{D})| \leqslant |A - |\nu|(\mathcal{D}')| + ||\nu|(\mathcal{D}') - |f|M(\mathcal{D}')|$$

$$< \epsilon + 2q\epsilon.$$

All that remains to complete the proof is to indicate where $\mathcal{D}'$ comes from. It arises from splitting intervals of $\mathcal{D}$ along the boundaries of the intervals of $\mathcal{K}$. This idea was used for the special case $\mathcal{K} = \{G, H\}$ in the proof of additivity in Section 2.4. The crucial property of the gauge is that $\gamma(z)$ intersect the boundary of an interval of $\mathcal{K}$ only when z belongs to that boundary. Such a γ entails this: when $zJ \in \mathcal{D}$ and J intersects some K in $\mathcal{K}$, $J \subseteq K$ unless z belongs to the boundary of K.

Figure 2 illustrates the situation in $\mathbf{R}^2$. In Figure 2(a), (b), and (c) the boundaries of the intervals of $\mathcal{K}$ are solid lines. A single interval J of $\mathcal{D}$ is shown. It is outlined by dashed lines. The dot in J is the tag of J. The locations of the tag in (a) and (b) are ruled out by a gauge of the type just described. Only (c) can occur. In this case J gives rise to three members of $\mathcal{D}'$. In two dimensions it is never necessary to split J into more than 4 intervals since no more than 4 members of $\mathcal{K}$ can have a common point.

In $\bar{\mathbf{R}}^p$ an interval J of $\mathcal{D}$ splits into no more than 2^p members of $\mathcal{D}'$, all having the same tag as J. Then $\mathcal{D}'$ is

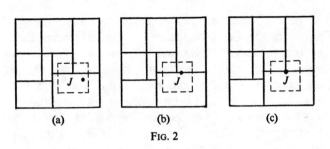

Fig. 2

γ-fine. The divisions $\mathcal{D}$ and $\mathcal{D}'$ also yield the same Riemann sum for any function.

S3.9. Iterated limits. The objective of this section is to formulate in general terms the notion of interchange of limits which lies behind the theorems on term-by-term integration of sequences and series. Once the concepts are stated in general terms, an iterated limits theorem of broad applicability can be stated and proved. This iterated limits theorem will be used in the next section in proving the monotone and dominated convergence theorems.

The first thing to be done is to define a general notion of limit which has all the kinds of limit we use as special cases. The discussion of the general notion of limit will be restricted to those facts which are directly useful. A full discussion may be found in McShane and Botts [9, p. 32 ff.].

Definition. Let X be a set. A *direction* $\mathcal{S}$ in X is a nonempty collection of nonempty subsets of X which is directed downward by inclusion. That is, for each S_1 and S_2 in $\mathcal{S}$ there is S_3 in $\mathcal{S}$ such that $S_3 \subseteq S_1$ and $S_3 \subseteq S_2$.

The notion of direction is at the heart of the limit concept. Some familiar limits are associated with the directions cited in the next paragraph.

The following are easily shown to be examples of directions. (a) Let X be the set of positive integers. Let $\mathcal{S}$ be the collection of all sets $\{n, n + 1, n + 2, \dots \}$ for $n = 1, 2, 3, \dots$. (b) Let $X = \mathbf{R}$. Fix $c \in \mathbf{R}$. Let $\mathcal{S}_c$ be the collection of deleted neighborhoods of c. That is, $S \in \mathcal{S}_c$ when $S = (u, c) \cup (c, v)$ for some u and v satisfying $u < c < v$. (c) Let I be a closed interval. Let X be the set of all tagged divisions of I. Let $S \in \mathcal{S}$ when S is the set of all γ-fine divisions $\mathcal{D}$ for some gauge γ on I.

A direction $\mathcal{S}$ in X serves to define a limit of functions with domain X and values in $\mathbf{R}^q$.

Definition. A is *the limit of f according to $\mathcal{S}$* provided that there exists S_ϵ in $\mathcal{S}$ such that $|A - f(x)| < \epsilon$ for all x in S_ϵ. Notation:

$$A = \lim_{\mathcal{S}} f \quad \text{or} \quad A = \lim_{x, \, \mathcal{S}} f(x).$$

The directions (a), (b), and (c) give familiar limits. The limit for (a) is the limit of sequences. That for (b) is $\lim_{x \to c} f(x)$ in the usual sense. Direction (c) gives the generalized Riemann integral as previously defined.

The uniqueness of limit which is tacitly included in the definition is easily proved by the argument used in the uniqueness theorem of the integral in Section 1.2. There is also a Cauchy criterion. That is, $\lim_{x, \, \mathcal{S}} f(x)$ exists if and only if there exists S_ϵ in $\mathcal{S}$ such that $|f(x_1) - f(x_2)| < \epsilon$ for all x_1 and x_2 in S_ϵ. The proof is a transcription of the one called for in Exercise 3 of Section 2.2.

Now suppose sets X and Y are given with directions $\mathcal{S}$ in X and $\mathcal{T}$ in Y. Let $f : X \times Y \to \mathbf{R}^q$ be given. Iterated limits are possible under these conditions. That is, it is meaningful to ask about existence of

$$\lim_{x, \, \mathcal{S}} \left(\lim_{y, \, \mathcal{T}} f(x, y) \right)$$

and

$$\lim_{y,\ \mathfrak{I}}\left(\lim_{x,\ \mathfrak{S}} f(x, y)\right).$$

Moreover the notion of uniform convergence of a sequence of functions is easily generalized to this situation.

Definition. $\lim_{x,\ \mathfrak{S}} f(x, y) = h(y)$ *uniformly for all* y *in* Y provided there exists S_ϵ in $\mathfrak{S}$ such that $|f(x, y) - h(y)| < \epsilon$ for all $y \in Y$ and all x in S_ϵ.

ITERATED LIMITS THEOREM. *Suppose* $\lim_{y,\ \mathfrak{I}} f(x, y)$ *exists for all* x *in* X *and* $\lim_{x,\ \mathfrak{S}} f(x, y) = h(y)$ *uniformly for all* y *in* Y. *Then the iterated limits exist and are equal. That is,*

$$\lim_{x,\ \mathfrak{S}}\left(\lim_{y,\ \mathfrak{I}} f(x, y)\right) = \lim_{y,\ \mathfrak{I}}\left(\lim_{x,\ \mathfrak{S}} f(x, y)\right).$$

Proof. For convenience let $g(x) = \lim_{y,\ \mathfrak{I}} f(x, y)$. The Cauchy criterion is the only tool in sight to use to show that $\lim_{x,\ \mathfrak{S}} g(x)$ exists. Using the uniformity of the limit on x, fix S_ϵ in $\mathfrak{S}$ so that $|f(x, y) - h(y)| < \epsilon$ for all x in S_ϵ and all y in Y. Now let x_1 and x_2 be in S_ϵ. Our objective is to show that $|g(x_1) - g(x_2)| < 4\epsilon$. This conclusion will come from the inequality

$$|g(x_1) - g(x_2)| \leqslant |g(x_1) - f(x_1, y)| + |f(x_1, y) - h(y)|$$
$$+ |h(y) - f(x_2, y)| + |f(x_2, y) - g(x_2)|.$$

By the definition of the $\mathfrak{I}$-limit, there are T_1 and T_2 in $\mathfrak{I}$ such that $|f(x_i, y) - g(x_i)| < \epsilon$ when $y \in T_i$ for $i = 1$ and $i = 2$. There is T in $\mathfrak{I}$ such that $T \subseteq T_1$ and $T \subseteq T_2$. Since T is not empty there is y in T which can be used in the inequality which closes the preceding paragraph. Thus the first and fourth terms of the right-hand side of the inequality are each less than ϵ. The second and third are each less than ϵ by the choice of S_ϵ. Thus $\lim_{x,\,S} g(x)$ exists by the Cauchy criterion. Call it A for convenience.

Now we must show that $A = \lim_{y,\,\mathfrak{T}} h(y)$ too. Fix x in S_ϵ so that $|A - g(x)| < \epsilon$ by using the fact that S is directed downward by inclusion. For this x there is T_x in $\mathfrak{T}$ for which $|g(x) - f(x,y)| < \epsilon$ when $y \in T_x$ since $g(x) = \lim_{y,\,\mathfrak{T}} f(x,y)$. Since

$$|A - h(y)| \leqslant |A - g(x)| + |g(x) - f(x,y)|$$
$$+ |f(x,y) - h(y)|$$

we conclude that $|A - h(y)| < 3\epsilon$ when $y \in T_x$. This completes the proof.

The last paragraph of this proof is a restatement in another notation of a part of the proof of the uniform convergence theorem. Indeed the whole proof is only a generalization of that proof with those parts of the former proof which relied on inequalities for integrals replaced by a slightly more elaborate argument based on the triangle inequality.

Some of the details in the solution of Exercise 2 above can be abbreviated by using the iterated limits theorem.

The iterated limits theorem will be used in Chapter 7 as well as in the next section.

S3.10. Proof of the monotone and dominated convergence theorems.

The discussion falls into several distinct parts signaled by (a), (b), etc.

(a) The statements of the monotone and dominated convergence theorems in Section 3.5 and 3.6 allow the interval I to be unbounded and to have any dimension. To get small majorants of sums over divisions of such an interval I it is helpful to extend to higher dimensions a tool used in the proof of the fundamental theorem in

Section S1.9. That tool is a function g which is everywhere positive on $\overline{\mathbf{R}}^p$ and which carries with it a gauge γ_g such that $gM(\mathfrak{D}) \leqslant 1$ when $\mathfrak{D}$ is any partial division of $\overline{\mathbf{R}}^p$ which is γ_g-fine.

Let G be continuous on $\overline{\mathbf{R}}$ and differentiable on $\mathbf{R}$ with $G'(x) > 0$ for all x in $\mathbf{R}$. Define g on $\mathbf{R}^p$ by setting $g(x) = G'(x_1)G'(x_2) \cdots G'(x_p)$ when $x = (x_1, x_2, \ldots, x_p)$. Define τ on the closed intervals in $\overline{\mathbf{R}}^p$ by setting $\tau(J) = 2^p \Delta G(J_1) \Delta G(J_2) \cdots \Delta G(J_p)$ when $J = J_1 \times J_2 \times \cdots \times J_p$. (As in earlier uses of this notation, $\Delta G([u, v]) = G(v) - G(u)$.)

The corollary to the straddle lemma in Section S1.9 provides an interval $\gamma(z)$ in $\mathbf{R}$ such that $G'(z)(v - u) < 2\Delta G([u, v])$ when $z \in [u, v]$ and $[u, v] \subseteq \gamma(z)$. When $z \in \mathbf{R}^p$ set $\gamma_g(z) = \gamma(z_1) \times \gamma(z_2) \times \cdots \times \gamma(z_p)$. Then $g(z) \cdot M(J) < \tau(J)$ when $z \in J$ and $J \subseteq \gamma_g(z)$. When z is an infinite point in $\overline{\mathbf{R}}^p$ the inequality $g(z)M(J) < \tau(J)$ holds for any J containing z since $M(J) = 0$ and $\tau(J) > 0$. Set $\gamma_g(z) = \overline{\mathbf{R}}^p$ in this case.

It is easy to check additivity of τ. There is no harm in supposing $\tau(\overline{\mathbf{R}}^p) = 1$ also. Thus $\tau(\mathfrak{D}) \leqslant 1$ when $\mathfrak{D}$ is a partial division of $\overline{\mathbf{R}}^p$. When $\mathfrak{D}$ is γ_g-fine $gM(\mathfrak{D}) \leqslant \tau(\mathfrak{D}) \leqslant 1$ as desired.

This sort of function will be useful in Section 6.4 in proving Fubini's theorem also.

(b) The monotone convergence theorem is an "if and only if" assertion. We can dispose of one of the two claims very easily. Suppose $\int_I f$ exists. Then $\int_I f_n \leqslant \int_I f_{n+1} \leqslant \int_I f$ for all n when f_n increases to f. Thus the sequence of integrals increases to a finite limit. All inequalities are reversed for a decreasing sequence. Thus the "only if" assertion is true.

In the remaining part of the monotone convergence theorem and in the dominated convergence theorem we wish to show that $\int_I f$ exists and equals $\lim_{n\to\infty}\int_I f_n$. Our common approach is to show that convergence of $f_n M(\mathfrak{D})$ to $\int_I f_n$ is uniform in n. This means that a gauge γ must be found such that $|\int_I f_n - f_n M(\mathfrak{D})| < \epsilon$ for all γ-fine $\mathfrak{D}$ and all n. Once this is done the iterated limits theorem tells us that

$$\lim_{n\to\infty}\Big(\lim_{\mathfrak{D}} f_n M(\mathfrak{D})\Big) = \lim_{\mathfrak{D}}\Big(\lim_{n\to\infty} f_n M(\mathfrak{D})\Big),$$

i.e., that $\lim_{n\to\infty}\int_I f_n = \int_I f$.

(c) The definition of γ does not depend on how f_n converges to f. The first step is to fix a gauge γ_n so that $|\int_I f_n - f_n M(\mathfrak{D})| < \epsilon/2^n$ when $\mathfrak{D}$ is a γ_n-fine division of I for $n = 1, 2, 3, \ldots$. Also fix a large integer N and a positive function g and its associated gauge γ_g as described in part (a) above. (The exact choice of N will be made later. It will depend on the nature of the convergence of f_n.) Let $n \geqslant N$. Let F_n be the set of all z such that $|f_k(z) - f(z)| \leqslant (\epsilon/4)g(z)$ when $k \geqslant n$. Then $F_{n-1} \subseteq F_n$ and $\bigcup_{n=N}^{\infty} F_n = I$. Set $E_N = F_N$ and $E_n = F_n - F_{n-1}$ when $n > N$. Some of the sets E_n may be empty but each z in I belongs to one and only one set E_n. Now set

$$\gamma(z) = \gamma_g(z) \cap \gamma_1(z) \cap \cdots \cap \gamma_n(z)$$

when $z \in E_n$. This defines γ on I. We must show that $|\int_I f_n - f_n M(\mathfrak{D})| < \epsilon$ when $\mathfrak{D}$ is γ-fine no matter what n may be.

(d) When $\mathfrak{D}$ is γ-fine it is also γ_n-fine for $n \leqslant N$. Thus $|\int_I f_n - f_n M(\mathfrak{D})| < \epsilon/2^n \leqslant \epsilon$ when $1 \leqslant n \leqslant N$. The challenging case is $n > N$.

For convenience set $\nu_n(J) = \int_J f_n$ for $n = 1, 2, 3, \ldots$. Define subsets of $\mathfrak{D}$ as follows: $\mathfrak{D}_i = \{zJ \in \mathfrak{D} : z \in E_i\}$, $\mathcal{E} = \bigcup_{i=N}^{n-1}\mathfrak{D}_i$, and $\mathcal{F} = \bigcup_{i\geqslant n}\mathfrak{D}_i$. ($\mathcal{E}$ and $\mathcal{F}$ depend on n but it is not essential to reflect that fact in the notation.)

When $\mathcal{D}$ is γ-fine each $\mathcal{D}_i$ is γ_1-fine, γ_2-fine, ..., γ_i-fine. Thus $\mathcal{F}$ is γ_n-fine but $\mathcal{E}$ is not. Consequently, when $n > N$,

$$\left| \int_I f_n - f_n M(\mathcal{D}) \right| \leq |\nu_n(\mathcal{E}) - f_n M(\mathcal{E})|$$
$$+ |\nu_n(\mathcal{F}) - f_n M(\mathcal{F})|$$
$$\leq |\nu_n(\mathcal{E}) - f_n M(\mathcal{E})| + \epsilon/2^n$$

according to Henstock's lemma applied to f_n and γ_n.

Now the auxiliary terms $\nu_i(\mathcal{D}_i)$ and $f_i M(\mathcal{D}_i)$, $N \leq i < n$, allow us to claim that

$$|\nu_n(\mathcal{E}) - f_n M(\mathcal{E})| \leq \sum_{i=N}^{n-1} |\nu_n(\mathcal{D}_i) - \nu_i(\mathcal{D}_i)|$$
$$+ \sum_{i=N}^{n-1} |\nu_i(\mathcal{D}_i) - f_i M(\mathcal{D}_i)|$$
$$+ \sum_{i=N}^{n-1} |f_i M(\mathcal{D}_i) - f_n M(\mathcal{D}_i)|.$$

Since $\mathcal{D}_i$ is γ_i-fine, Henstock's lemma yields the majorant $\sum_{i=N}^{n-1} \epsilon/2^i$ for the middle sum on the right. Since the tags of $\mathcal{D}_i$ are in E_i and $n > i$, the definition of E_i allows us to say that $|f_i M(\mathcal{D}_i) - f_n M(\mathcal{D}_i)| \leq (\epsilon/4) g M(\mathcal{D}_i)$. Thus the last sum on the right has the majorant $(\epsilon/4) g M(\mathcal{E})$. This is less than $\epsilon/4$ since $\mathcal{E}$ is also γ_g-fine. We may assume that $\sum_{i=N}^{n-1} \epsilon/2^i < \epsilon/4$ without harm. Thus the findings of this paragraph may be summarized as

$$|\nu_n(\mathcal{E}) - f_n M(\mathcal{E})| \leq \sum_{i=N}^{n-1} |\nu_n(\mathcal{D}_i) - \nu_i(\mathcal{D}_i)| + \epsilon/4 + \epsilon/4.$$

Carrying this back into the conclusion of the preceding

paragraph, we have

$$\left| \int_I f_n - f_n M(\mathcal{D}) \right| < \sum_{i=N}^{n-1} |\nu_n(\mathcal{D}_i) - \nu_i(\mathcal{D}_i)| + 3\epsilon/4$$

when $n > N$.

(e) At this point the remaining task is to choose N so that the sum on the right is less than $\epsilon/4$. For the first time it is advantageous to bring into play the nature of the convergence of f_n to f.

When the sequence is monotone all the differences $\nu_n(J) - \nu_i(J)$, i.e., $\int_J (f_n - f_i)$, are of the same sign and $|\nu_n(J) - \nu_i(J)| \leqslant |\int_J (f_n - f_N)|$ when $N \leqslant i < n$. Consequently

$$\sum_{i=N}^{n-1} |\nu_n(\mathcal{D}_i) - \nu_i(\mathcal{D}_i)| \leqslant \sum_{J \in \mathcal{E}} \left| \int_J (f_n - f_N) \right| \leqslant \left| \int_I (f_n - f_N) \right|.$$

The sequence of integrals $\int_I f_n$ is convergent by hypothesis. Thus N can be chosen so that $|\int_I f_n - \int_I f_N| < \epsilon/4$ when $n > N$. Consequently $|\int_I f_n - f_n M(\mathcal{D})| < \epsilon$ when $n > N$. This completes the argument for monotone sequences.

(f) We take now the hypothesis of the dominated convergence theorem in the form $|f_n - f_m| \leqslant h$ for all m and n with h integrable over I. An auxiliary sequence of functions will be helpful. First note that, for given j and k, $\bigvee_{j \leqslant m \leqslant n \leqslant k} |f_n - f_m|$ is an integrable function which does not exceed h. Moreover it increases to a finite limit function t_j when j is held fixed and k tends to ∞. The monotone convergence theorem tells us that t_j is also integrable on I. A second look at the definition of t_j shows that $t_j \geqslant t_{j+1}$ for all j and that $\lim_{j \to \infty} t_j(x) = 0$ for all x. A second application of the monotone convergence theorem asserts that $\int_I t_j$ tends to zero. Thus we can choose N so that $\int_I t_N < \epsilon/4$.

Since $|f_n - f_i| \leqslant t_N$ when $N \leqslant i < n$,

$$\sum_{i=N}^{n-1} |\nu_n(\mathcal{D}_i) - \nu_i(\mathcal{D}_i)| \leqslant \sum_{J \in \mathcal{S}} \int_J t_n \leqslant \int_I t_N < \epsilon/4.$$

This completes the argument for the dominated convergence theorem since we again have $|\int_I f_n - f_n M(\mathcal{D})| < \epsilon$ when $n > N$.

3.11. Exercises.

4. Let f be integrable on the unbounded interval I. Suppose $|f M(\mathcal{D}) - \int_I f| < \epsilon$ for every γ-fine division $\mathcal{D}$ of I. Let $\mathcal{S}$ be the subset of $\mathcal{D}$ consisting of unbounded intervals. Estimate $\sum_{zJ \in \mathcal{S}} |\int_J f|$.

5. Let $f(x) = x \sin 1/x$ when $x \neq 0$ and $f(0) = 0$. Show that f' is integrable on $[0, 1]$ but $|f'|$ is not.

6. Let f be real-valued. Show that integrability of any two of f, $|f|$, f^+, and f^- implies integrability of all four.

7. Suppose I is unbounded. Give a proof not using the dominated convergence theorem that $\lim_{n \to \infty} \int_I f_n = \int_I f$ when f_n converges uniformly to f on each bounded interval and $|f_n| \leqslant g$ with g integrable on I.

8. Show that $F : [a, b] \to \mathbf{R}^q$ is a function of bounded variation if and only if all its real component functions are functions of bounded variation.

9. Let $f : [a, b] \to \mathbf{R}$ be continuous and piecewise monotone, i.e., monotone on each subinterval of some division of $[a, b]$. Show that the graph of f is a curve of finite length.

10. Show that the circular arc $y = \sqrt{1 - x^2}$, $a \leqslant x \leqslant 1$, has length given by $\int_a^1 (1/\sqrt{1 - x^2}) \, dx$.

11. Let $f(x) = \sum_{n=0}^{\infty} e^{-nx} \cos nx$ for $x > 0$. Let $t > 0$. Show that $\int_t^{\infty} f(x)\, dx$ can be obtained by term-by-term integration of the series. Is f integrable on $[0, \infty]$?

12. Give an alternative proof of the monotone convergence theorem by showing directly from the definition of the integral that $A = \int_I f$ where $A = \lim_{n \to \infty} \int_I f_n$. Hint: use the gauge γ defined in part (c) of Section S3.10.

13. Give an example which shows that existence of an integrable function h such that $|f_n - f_m| \leqslant h$ is not necessary for $f_n M(\mathcal{D})$ to converge to $\int_I f_n$ uniformly with respect to n.

14. Construct functions f_n which converge to a function f on I so that $\int_I f_n$ converges to $\int_I f$ but $\int_J f_n$ does not converge to $\int_J f$ for some subinterval J.

15. Suppose $\lim_{n \to \infty} f_n = f$ on I and $f_n M(\mathcal{D})$ converges to $\int_I f_n$ uniformly with respect to n. Show that $\lim_{n \to \infty} \int_J f_n = \int_J f$ uniformly for all subintervals J.

16. Let $\lim_{n \to \infty} f_n = f$ uniformly on a bounded interval I. Show that $f_n M(\mathcal{D})$ converges to $\int_I f_n$ uniformly with respect to n. (Recall that it has already been shown that $f_n M(\mathcal{D})$ converges to $f M(\mathcal{D})$ uniformly for all divisions $\mathcal{D}$.)

CHAPTER 4

INTEGRATION ON SUBSETS OF INTERVALS

Many problems require the notion of integration over sets in $\mathbf{R}^p$ which are not intervals. The calculation of area, volume, moments, and work brings this idea to the attention of the calculus student. As was pointed out in Section 1.6, one way to define $\int_E f$ when f is defined initially on E is to extend f by setting it equal to zero outside E and integrate the extension over some closed interval containing E, say $\overline{\mathbf{R}}^p$. The obvious advantage of this method is that the facts for integration over intervals are available for the development of the properties of the integral over noninterval sets. The results of Chapters 2 and 3 are important tools in this chapter.

The opening section takes up a special question which is closely related to the main topic of the chapter. The question is this: What properties of f make it possible that $\int_J f = 0$ for every closed subinterval J of I? It should be no surprise that the set on which f takes nonzero values plays a special role. Such a set has to be sparsely spread through I. These sets are called null sets since they are negligible in integration.

Section 4.2 applies the idea of null set and the facts about null sets to the monotone and dominated convergence theorems. More general versions of these two theorems are given in which it is not supposed that $\lim_{n\to\infty} f_n(x)$ exists for every x.

The properties of $\int_E f$ which relate to finite combinations of sets formed by the set operations of difference, union, and intersection are developed in Section 4.3. The facts from Chapter 2 are heavily used here.

Section 4.4 addresses the question of integrability of a function which is continuous on a closed and bounded set. Such two-dimensional sets as those bounded by two graphs of continuous functions are important special cases. The monotone convergence theorem has an important place in the proof of the main result.

In Section 4.5 the relation of the integral over the union of a sequence of sets to the integrals over individual sets in the sequence is developed for two types of sequences. The monotone and dominated convergence theorems are the essential facts here. A more general version of the comparison test for absolute integrability is one of the subsidiary conclusions in this section.

Area in two dimensions and volume in three dimensions are formulated as special cases of the notion of measure of subsets of $\mathbf{R}^p$ in Section 4.6. The measure of E is defined as the integral over E of the constant function with value one. Consequently, the results of the preceding sections can be specialized to develop the properties of the measure function and the collection of sets which have a finite measure. The sets having finite measure are a subclass of the sets whose intersections with bounded intervals possess a measure. The basic properties of this larger class, called the measurable sets, are also brought out.

4.1. Null functions and null sets. The function which is zero everywhere on I has zero as the value of its integral on every subinterval of I. We saw very early in Chapter 1 that a function f can have nonzero values at many points in an interval in $\overline{\mathbf{R}}$ and still have a zero integral on every subinterval. The same behavior can occur on higher dimensional intervals. Such functions play a significant role in the theory of the integral. Thus we shall examine their properties.

The first fact to note is that $\int_J f = 0$ *for every subinterval of I if and only if* $\int_I |f| = 0$. The implication one way follows from the inequalities $|\int_J f| \leqslant \int_J |f| \leqslant \int_I |f|$. The reverse implication can be proved from Henstock's lemma by the argument used in Example 1 of Section 3.1. The choice of one of these equivalent properties for the attachment of a name is based purely on convenience.

Definition. A function $f : I \to \mathbf{R}^q$ is a *null function* on I provided $\int_I |f| = 0$.

Suppose functions f and g are given and their difference $f - g$ is a null function on I. It is easy to argue that both are integrable on I if one of them is integrable and that $\int_J f = \int_J g$ for every subinterval J. Consequently one may say that addition of a null function to a given function has no effect on integrability or the value of the integral. It must not be overlooked that functions which differ by a null function may be very different in other respects, however. Recall the function which is zero on the rationals and one on the irrationals. It is continuous nowhere. It differs from the constant function whose value is one by a null function. The constant function is continuous everywhere, of course.

Now let's examine null functions on the basic level of Riemann sums and gauges. From the definition of $\int_I |f|$, f is a null function on I provided $|f| M(\mathcal{D}) < \epsilon$ for every division $\mathcal{D}$ of I which is γ-fine for a suitable γ. Clearly no restriction need be placed on $\gamma(z)$ when $f(z) = 0$ since the corresponding term $|f(z)| M(J)$ is zero whatever J may be.

This observation directs our attention away from the whole interval I and toward one of its subsets and partial divisions whose tags lie in that subset. The following definition captures the essential idea.

Definition. Let f be defined at least on the subset E of $\overline{\mathbf{R}}^p$. The set E is *fM-null* provided that to each ϵ there corresponds a gauge γ on E such that $|f|M(\mathcal{S}) < \epsilon$ for every γ-fine partial division $\mathcal{S}$ whose tags lie in E.

(Recall that a partial division is a finite set of nonoverlapping intervals and that this is equivalent to its being a subset of a division of $\overline{\mathbf{R}}^p$ or of any other closed interval which contains all the given intervals.)

So far, we have merely restated the idea of null function. The gain in the restatement is that a few easily proved properties of this new concept are quite fruitful. We now turn to the three properties which will be most helpful.

It is clear from the definition that *any subset of an fM-null set is also fM-null.*

We shall show next that *any countable union of fM-null sets is also fM-null.* Let $E = \bigcup_{i=1}^{\infty} E_i$ and suppose each E_i is *fM*-null. These sets need not be pairwise disjoint. Set $F_1 = E_1$, $F_2 = E_2 - E_1$, $F_3 = E_3 - (E_1 \cup E_2)$, and so on. The sets $F_1, F_2, F_3, \ldots$ are pairwise disjoint and have E as their union. Each F_i is *fM*-null since $F_i \subseteq E_i$. Let γ_i be defined on F_i so that $|f|M(\mathcal{S}) < \epsilon/2^i$ when $\mathcal{S}$ is a γ_i-fine partial division whose tags are in F_i. Define γ on E by setting $\gamma(z) = \gamma_i(z)$ when $z \in F_i$. Let $\mathcal{S}$ be γ-fine. Let $\mathcal{S}_i$ be the subset of $\mathcal{S}$ whose tags lie in F_i. Ignore all $\mathcal{S}_i$ which are empty. Otherwise $\mathcal{S}_i$ is a γ_i-fine partial division. Consequently $|f|M(\mathcal{S}_i) < \epsilon/2^i$. Since $|f|M(\mathcal{S}) = \sum_i |f|M(\mathcal{S}_i)$ we conclude that $|f|M(\mathcal{S}) < \sum_{i=1}^{\infty} \epsilon/2^i$. That is, E is *fM*-null.

A third useful fact is that *E is gfM-null for any g when E is fM-null.* (Only the obvious requirement that gf be meaningful need be met.) The proof of this property uses the two preceding properties. Set $E_i = \{x \in E : |g(x)| \leqslant i\}$. Then $E = \bigcup_{i=1}^{\infty} E_i$ and it suffices to show that each E_i is gfM-null. Since E_i is fM-null there is a gauge γ_i on E_i such that $|f|M(\mathcal{S}) < \epsilon / i$ when $\mathcal{S}$ is γ_i-fine and has tags in E_i. Consequently $|gf|M(\mathcal{S}) \leqslant i|f|M(\mathcal{S}) < \epsilon$ when $\mathcal{S}$ is γ_i-fine and has its tags in E_i.

Now we have the working tools we need to examine the extent to which the particular characteristics of f determine whether a set E is fM-null.

The function f which is one everywhere plays a special role. Since the sum $fM(\mathcal{S})$ is the same as $M(\mathcal{S})$ when f equals one everywhere, we may just as well say E *is M-null* when E is fM-null and f is constant with value one.

Exercise 1. Let $f : I \to \mathbf{R}^q$ be given. Let $E = \{x \in I : f(x) \neq 0\}$. Suppose E is M-null. Show that E is fM-null and that $\int_I |f| = 0$.

There is also a converse to the proposition in Exercise 1 which further clarifies the relation of M-null sets to null functions.

Exercise 2. Suppose that $\int_I |f| = 0$. Show that

$$\{x \in I : f(x) \neq 0\}$$

is M-null.

These two exercises characterize a null function on I as a function which vanishes outside an M-null subset of I. If we want to know more about null functions we must learn more about the M-null sets.

We already know from Example 4, page 19, that any countable subset of a bounded interval in $\mathbf{R}$ is L-null. Essentially the same argument as the one used there shows that each countable subset of $\overline{\mathbf{R}}^p$ is M-null. What other sets are M-null? It appears that the points of a set E must be sparse in some sense if E is to be M-null. Surely E cannot contain a nondegenerate bounded closed interval K because K is not M-null. To see that K is not M-null we need note only that each gauge γ on K has a γ-fine division $\mathcal{D}$ of K and that $M(\mathcal{D}) = M(K)$.

We continue our catalog of M-null sets with the most obvious candidates.

Example 1.   Show that any degenerate interval in $\overline{\mathbf{R}}^p$ is M-null.

Solution.   It is convenient to treat special cases. Suppose I is a bounded degenerate interval. Select a bounded open interval G and a bounded closed interval H so that $I \subseteq G \subseteq H$ and $M(H) < \epsilon$. Let $\gamma(z) = G$ for all z in I. When $\mathcal{E}$ is a γ-fine partial division there is a division $\mathcal{D}$ of H such that $\mathcal{E} \subseteq \mathcal{D}$. Consequently $M(\mathcal{E}) \leqslant M(\mathcal{D}) = M(H) < \epsilon$.

Next let I be a degenerate interval made up entirely of points at infinity. Since M vanishes on all unbounded intervals, $M(\mathcal{E}) = 0$ when all tags of E are infinite points. Thus I is M-null.

Any degenerate interval I in $\mathbf{R}^p$ is expressible as a countable union of bounded degenerate intervals. Thus I is M-null, since it is a countable union of M-null sets.

Finally, any degenerate interval in $\overline{\mathbf{R}}^p$ is the union of a degenerate interval in $\mathbf{R}^p$ and a set of points at infinity. Thus it is M-null.

Next we consider the graphs of real-valued functions. A function $f : [a, b] \to \mathbf{R}$ has for its graph a subset of $\mathbf{R}^2$ which seems likely to be M-null, at least under some smoothness restrictions on f. When the domain of f is an interval in $\mathbf{R}^2$ the graph is a surface in $\mathbf{R}^3$ which we conjecture to be M-null when the three-dimensional interval measure is used. The next exercise proposes the general case.

Exercise 3. Let I be a closed, bounded interval in $\mathbf{R}^{p-1}$ with $p \geqslant 2$. Let $f : I \to \mathbf{R}$ be continuous. Let E be the graph of f, i.e., $E = \{(x, y) : x \in I \text{ and } y = f(x)\}$. Show that E is an M-null subset of $\mathbf{R}^p$.

Since any interval in $\mathbf{R}^{p-1}$ is a countable union of closed, bounded intervals it follows from Exercise 3 that the graph of a function which is continuous on any type of interval in $\mathbf{R}^{p-1}$ is M-null in $\mathbf{R}^p$.

Other examples of M-null sets can be created by taking subsets and countable unions of graphs of continuous functions.

The remainder of this section is devoted to some comments.

The first M-null sets we found were the countable sets. Now we have the means to show that M-null sets need not be countable. When $p \geqslant 2$ there are degenerate intervals which are not countable. In $\mathbf{R}$ we must look elsewhere for an L-null set which is not countable. The construction of one well-known example, the Cantor set, is outlined in Exercise 21 at the end of this chapter.

Cantor's set enables us to make another point concerning the fundamental theorem. Recall that $\int_a^b f = F(b) - F(a)$ when $F'(x) = f(x)$ except possibly on a countable subset of $[a, b]$. Can the exceptional set be allowed to be L-null rather than countable? No, it cannot. The construction of a suitable counterexample is outlined in Exercise 22. The counterexample is a function $F : [0, 1] \to \mathbf{R}$ which is continuous on $[0, 1]$ and differentiable at all points which do not belong to the Cantor set. Furthermore $F'(x) = 0$ when x is not in the Cantor set. Thus $\int_0^1 f = 0$ when f is any function which agrees with F' at the points where F' is zero. The function F is constructed so that $F(0) = 0$ and $F(1) = 1$. Consequently, the conclusion of the fundamental theorem cannot hold.

Generalization of the fundamental theorem by allowing the exceptional set to be L-null while keeping all other assumptions the same is blocked. A tradeoff is possible, however. A stronger continuity condition on F compensates for the weaker requirement that $F'(x) = f(x)$ except possibly on an L-null set. The appropriate continuity concept is introduced in Exercise 24 at the end of the chapter. Exercise 25 calls for the proof of the modified fundamental theorem.

4.2. Convergence almost everywhere. Null sets and null functions make possible an increase in generality of the monotone and dominated convergence theorems. The more general versions are customarily expressed in the following terminology.

A statement in terms of points in a subset of $\overline{\mathbf{R}}^p$ is said to hold *almost everywhere* if it is true except possibly at the points of an M-null set. For instance, a null function on I vanishes almost everywhere on I. Alternately, one may say that a statement which mentions a specific variable, say x, is true for *almost all* x when it is true for all x in a set whose complement is M-null. Thus f is a null function on I when $f(x) = 0$ for almost all x in I.

The following version of the dominated convergence theorem is a typical use of null sets and the language just introduced. It may, of course, be expressed also in terms of series.

Suppose $\lim_{n\to\infty} f_n = f$ *almost everywhere in* I *and, for each* n, $|f_n| \leqslant h$ *almost everywhere on* I. *Suppose each* f_n *and* h *is integrable on* I. *Then* f *is integrable on* I *and* $\lim_{n\to\infty} \int_I f_n = \int_I f$.

To deduce this assertion from the one in Section 3.6 we form a single M-null set E by taking the union of those sets where $\lim_{n\to\infty} f_n(x) = f(x)$ fails and $|f_n(x)| \leqslant h(x)$ fails for $n = 1, 2, 3, \ldots$. On E replace the given values of f, h, and each f_n by zero, say. Then the modified functions will satisfy the hypotheses given in Section 3.6. They are equivalent to the given functions in integrability and the values of their integrals. Thus the assertion above follows from the one in Section 3.6.

The modification of the monotone convergence theorem which has been promised in Section 3.5 is more substantial. To get to it we must prove the following proposition.

Suppose $(f_n)_{n=1}^{\infty}$ *is an increasing sequence of nonnegative integrable functions. Suppose the sequence of integrals* $\int_I f_n$ *is bounded. Then* $\lim_{n\to\infty} f_n(x)$ *is finite for almost all* x *in* I.

One way to prove this proposition is to use the monotone convergence theorem itself. Let h be the function which equals one on the set E where $\lim_{n\to\infty} f_n(x) = \infty$ and equals zero elsewhere. The proof is complete when we show that $\int_I h = 0$. The monotone convergence theorem enters twice in showing that h is the limit of a decreasing sequence $(h_i)_{i=1}^{\infty}$.

Let $f(x) = \lim_{n\to\infty} f_n(x)$ when $x \in I - E$. For a fixed positive integer i the sequence $(i^{-1} f_n)_{n=1}^{\infty}$ also increases to ∞ on E but its limit is $i^{-1} f(x)$ when $x \in I - E$. This brings us closer to h. It is another step forward to cut off the graph of $i^{-1}f_n$ at height 1. That is, we form $g \wedge (i^{-1}f_n)$ where $g(x) = 1$ everywhere. These functions increase to a limit h_i and the sequence $(h_i)_{i=1}^{\infty}$ decreases to h. Since the integrals $\int_I f_n$ all have a bound A one shows readily that $\int_I h_i \leqslant i^{-1}A$. Then $\int_I h = 0$ follows from the monotone convergence theorem.

Exercise 4.   Supply the missing details of the above argument.

When $(f_n)_{n=1}^{\infty}$ is monotone, either $(f_n - f_1)_{n=1}^{\infty}$ or $(f_1 - f_n)_{n=1}^{\infty}$ is an increasing sequence of nonnegative functions. Thus we get the following theorem from the assertion above and the monotone convergence theorem given in Section 3.5.

Let $(f_n)_{n=1}^{\infty}$ be a monotone sequence of integrable functions. Let the sequence of integrals $(\int_I f_n)_{n=1}^{\infty}$ be bounded. Then $\lim_{n \to \infty} f_n(x)$ is finite for almost all x in I. If f is any function which equals $\lim_{n \to \infty} f_n(x)$ almost everywhere, f is integrable and $\int_I f = \lim_{n \to \infty} \int_I f_n$.

Instead of converting this theorem to series form it is more useful to pass on to a theorem about functions which may have values in $\mathbf{R}^q$ for any q.

Suppose each f_n is absolutely integrable and $\sum_{n=1}^{\infty} \int_I |f_n|$ is finite. Then $\sum_{n=1}^{\infty} f_n(x)$ converges for almost all x and $\int_I f = \sum_{n=1}^{\infty} \int_I f_n$ when $f = \sum_{n=1}^{\infty} f_n$ almost everywhere. Moreover f is absolutely integrable.

The series $\sum_{n=1}^{\infty} |f_n|$ converges almost everywhere, since its partial sums form an increasing sequence of nonnegative functions with bounded integrals. Absolute convergence implies convergence in $\mathbf{R}^q$. The integrability of $\sum_{n=1}^{\infty} |f_n|$ comes from the monotone convergence theorem. This series of absolute values is the majorant needed for application of the dominated convergence theorem to $\sum_{n=1}^{\infty} f_n$. Finally, integrability of $|f|$ follows from the comparison test since $|f| \leqslant \sum_{n=1}^{\infty} |f_n|$.

4.3. Integration over sets which are not intervals. Suppose $E \subseteq \overline{\mathbf{R}}^p$ but E is not known to be an interval. The integral of $f : E \rightarrow \mathbf{R}^q$ is defined as follows. Set $g(x) = f(x)$ when $x \in E$ and $g(x) = 0$ otherwise. Select an interval I containing E. The function f is integrable on E if and only if g is integrable on I. Set $\int_E f = \int_I g$ when $\int_I g$ exists.

This is a satisfactory definition provided two conditions are met. When E is contained in more than one interval I the choice of I must not alter integrability of f or the value of $\int_E f$. When E itself is an interval the value of $\int_I g$ must agree with the value of $\int_E f$ obtained from the Riemann sum definition applied to f on E. Each of these consistency conditions is readily deducible from the following assertion.

Example 2. Let I and J be intervals with $J \subseteq I$. Suppose $g(x) = 0$ outside J and $\int_J g$ exists. Show that $\int_I g$ exists and $\int_J g = \int_I g$.

Solution. Let G be the open interval obtained from J by removal of its faces. Then $J - G$ is a finite union of degenerate intervals. Consequently, $J - G$ is M-null and the function h which agrees with g on $J - G$ and is zero elsewhere is a null function. Thus $\int_J g = \int_J (g - h)$. To show that $\int_J (g - h) = \int_I (g - h)$ we examine Riemann sums for the two integrals. Since G is open, we may use only gauges γ satisfying $\gamma(z) \subseteq G$ when $z \in G$. Note that $g - h$ vanishes outside G. It is easy to see that each Riemann sum of $g - h$ for a γ-fine division of I equals a Riemann sum for a γ-fine division of J. Consequently, $\int_I (g - h)$ exists and equals $\int_J (g - h)$. But $\int_I h = 0$ also. Thus $\int_J g = \int_I g$.

The relation of the integral to some of the set operations is the first aspect of the integral over noninterval sets that needs study.

Suppose f is integrable on E and F. Suppose $E \subseteq F$. Then f is integrable on $F - E$ and $\int_{F-E} f = \int_F f - \int_E f$.

This is an immediate consequence of the definition and the linearity of the integral.

Finite additivity extends to noninterval sets, too. *Let f be integrable on E_i for $i = 1, 2, \ldots, n$. Suppose $E_i \cap E_j$ is M-null when $i \neq j$. Set $E = \bigcup_{i=1}^n E_i$. Then f is integrable on E and $\int_E f = \sum_{i=1}^n \int_{E_i} f$.*

This is easily proved as follows. Let g and g_i be the zero extensions of f outside E and E_i, respectively. Then $g(x) = \sum_{i=1}^n g_i(x)$ almost everywhere since $E_i \cap E_j$ is M-null when $i \neq j$. Linearity of the integral supplies the desired conclusion.

These two properties yield a useful observation concerning $\int_E f$ and $\int_F f$ when $E \subseteq F$ and $F - E$ is M-null. Existence of either $\int_E f$ or $\int_F f$ implies existence of the other integral and their equality since $\int_{F-E} f = 0$ no matter how f behaves on the M-null set $F - E$. Consequently it is a matter of indifference whether one speaks of integration over I' or I when I' and I are intervals which differ by an M-null set, e.g., $I' = \mathbf{R}^p$ and $I = \overline{\mathbf{R}}^p$.

There are a few other properties of the integral which carry over from intervals to more general sets. Linearity, component-by-component integration, and inequalities are all readily proved for the general case from the facts given in Section 2.1.

It is not being claimed that integrability of f on a given set implies integrability on every subset. This is not even true when f is constant. More about nonintegrability on subsets appears in the next paragraph and on page 127.

When f is integrable on E and on F it does not follow that f is integrable on $E \cap F$. Neither does it follow that f is integrable on $E \cup F$ unless $E \cap F$ is M-null, as in the proposition above. A stronger assumption on f is needed.

Suppose f is absolutely integrable on E and on F. Then f is absolutely integrable on $E \cap F$ and $E \cup F$.

By going from vectors to components to positive and negative parts we may assume f is nonnegative. Let g and h be the zero extensions of f outside E and F, respectively. Then $g \wedge h$ and $g \vee h$ are the zero extensions of f outside $E \cap F$ and $E \cup F$. Both of these are integrable.

Exercise 5. Produce an example which shows f cannot be assumed merely integrable in the proposition above.

The comparison test of Section 3.2 yields the following extension.

Suppose $E \subseteq I$. Let f be integrable on E. Let g be a nonnegative function which is integrable on I. If $|f| \leqslant g$ on E then $|f|$ is integrable on E.

For integration over noninterval sets to have much significance there must be interesting examples. With the properties above we can make limited headway in producing examples. For instance, we can claim integrability of f over any set which is expressible as a finite union of intervals provided f is integrable over intervals. In order to deal with larger classes of sets we must be able to deal with limits of sequences of functions. The monotone and dominated convergence theorems will serve us well in doing this.

4.4. Integration of continuous functions on closed, bounded sets.   Some standard terminology about sets will be useful. Let I be an interval in $\overline{\mathbf{R}}^p$ and let $E \subseteq I$. The set E is *open in* I provided each x in E belongs to an open interval $\gamma(x)$ such that $I \cap \gamma(x) \subseteq E$. E is *closed* when $\overline{\mathbf{R}}^p - E$ is open in $\overline{\mathbf{R}}^p$. E is *bounded* when E is contained in a bounded interval.

The previous terminology for open and closed intervals is compatible with these more general terms. Observe that (a, b) is open in any interval containing (a, b). On the other hand $[a, b)$ is open in $[a, b]$ but not in $[c, b]$ when $c < a$.

Our objective is to show that continuity of a function f defined on a closed and bounded set E suffices for integrability of f on E. Keep in mind that we actually deal with $\int_I g$ where g agrees with f on E and is zero elsewhere. Thus we must cope with some discontinuities. When g is nonnegative it is nevertheless the limit of a decreasing sequence of functions of simple structure.

Definition.   A function $f : I \to \mathbf{R}^q$ is a *step function* provided there is a partition of I into subintervals $I_1, I_2, \ldots, I_n$ such that f is constant on each interval I_k.

It should be emphasized that $I_j \cap I_k$ is empty when $j \neq k$ and that $I = \bigcup_{k=1}^{n} I_k$. These requirements cannot be met when all the intervals are closed and nondegenerate. A couple of simple examples may help to clarify this.

The function which equals one on $[0, 1)$ and zero at 1 is a step function on $[0, 1]$. In $\mathbf{R}^2$ let $f(x, y) = 1$ when $0 \leqslant x \leqslant 1$ and $0 \leqslant y \leqslant 2$. Let $f(x, y) = 0$ otherwise. This, too, is a step function since $\mathbf{R}^2$ can be broken up into 5 intervals so that f is constant on each one.

A function which is constant on a bounded interval and zero elsewhere is integrable on the whole space. Consequently, by the linearity of the integral, a step function f is integrable on I when I is bounded and also when I is unbounded provided f vanishes outside some bounded interval.

In solving the next exercise keep in mind the following well-known fact. Let E be closed and bounded. A function $f : E \to \mathbf{R}$ which is continuous on E assumes a maximum and a minimum value on E.

Exercise 6. Suppose I is a closed, bounded interval and E is a closed subset of I. Let $f : E \to \mathbf{R}$ be continuous and nonnegative on E. Set $f(x) = 0$ outside E. Construct a decreasing sequence $(f_n)_{n=1}^{\infty}$ of step functions so that $\lim_{n \to \infty} f_n(x) = f(x)$ for all x in I.

Now we are ready for the proof of the following. *Let E be a closed, bounded subset of $\mathbf{R}^p$. Suppose $f : E \to \mathbf{R}^q$ is continuous on E. Then $\int_E f$ exists.*

Note that continuity of f passes along to its real-valued component functions. From a real function f continuity also descends to the positive and negative parts f^+ and f^-. Since integrability passes backward along this same chain it is enough to give the proof for nonnegative f.

Let $f(x)$ be zero for x outside E. Fix a bounded interval I containing E and construct step functions which decrease to f on I. The step functions are integrable since I is bounded. The sequence of integrals has a finite limit since it is decreasing and bounded below by zero. Thus

$\int_I f$ exists by the monotone convergence theorem. This is the same as $\int_E f$.

Most of the examples of elementary multi-dimensional integral calculus fall under this proposition. For instance, in $\mathbf{R}^2$ the closed disk $|x| \leqslant r$, the closed triangular disk, and more generally the set $\{(x, y) : a \leqslant x \leqslant b$ and $g_1(x) \leqslant y \leqslant g_2(x)\}$ for bounded $[a, b]$ and continuous g_1 and g_2 are all closed and bounded sets. Thus a continuous function on a set of this type is integrable. The value of the integral is usually found by iterated integration, of course.

4.5. Integrals on sequences of sets. When $E = \bigcup_{n=1}^{\infty} E_n$ the integral of a function f over E can be related to its integrals over the sets E_n under appropriate restrictions. Two types of sequences $(E_n)_{n=1}^{\infty}$ are most useful. An *expanding* sequence has the property that $E_n \subseteq E_{n+1}$ for all n. The sequence of integrals $\int_{E_n} f$ has $\int_E f$ as its limit when f is suitably restricted. The sequence $(E_n)_{n=1}^{\infty}$ consists of *nonoverlapping* sets when $E_i \cap E_j$ is M-null for $i \neq j$. The integrals $\int_{E_n} f$ form the terms of a series whose sum is $\int_E f$ for appropriate f when the sets are nonoverlapping.

An expanding sequence has a companion sequence of nonoverlapping sets and vice versa. In principle, it is enough to deal with only one of the two types. However, some types of hypotheses lend themselves more readily to one formulation than to the other. Both will be used.

The monotone convergence theorem provides a simple proof of the following.

Let f be nonnegative on the union E of the expanding sequence $(E_n)_{n=1}^{\infty}$. Let f be integrable on each E_n. Then $\int_E f$ exists if and only if $\lim_{n \to \infty} \int_{E_n} f$ is finite. Moreover $\int_E f = \lim_{n \to \infty} \int_{E_n} f$.

From integrability of nonnegative functions it is a fairly small step to absolute integrability. By using a sequence of nonoverlapping sets we arrive at a generalization of finite additivity which is called *countable additivity*.

Let $(E_n)_{n=1}^{\infty}$ be a sequence of nonoverlapping sets. Suppose f is absolutely integrable on each E_n. Let $E = \bigcup_{n=1}^{\infty} E_n$. Then f is absolutely integrable on E if and only if $\sum_{n=1}^{\infty} \int_{E_n} |f|$ is finite. Moreover $\int_E f = \sum_{n=1}^{\infty} \int_{E_n} f$.

Let f_n agree with f on E_n and vanish elsewhere. When $\sum_{n=1}^{\infty} \int_I |f_n|$ is finite all the conclusions concerning f follow from the final theorem of Section 4.2. (See p. 112.) Conversely, when $|f|$ is integrable on E the convergence of $\sum_{n=1}^{\infty} \int_I |f_n|$ follows from the fact that $|f|$ is a dominant of all partial sums of $\sum_{n=1}^{\infty} |f_n|$.

Bounded sets are often convenient choices for the sets E_n in the above criteria for integrability. This is because continuous functions, among others, are integrable over bounded sets.

Exercise 7. In $\mathbf{R}^p$ set $H_n = [-n, n] \times [-n, n] \times \cdots \times [-n, n]$ for $n = 1, 2, 3, \ldots$. Suppose f is integrable on each H_n. Suppose $|f(x)| \leqslant b_1$ when $x \in H_1$ and $|f(x)| \leqslant b_n$ when $x \in H_n - H_{n-1}$ for $n \geqslant 2$. Suppose also that $\sum_{n=1}^{\infty} n^{p-1} b_n$ is finite. Show that f is absolutely integrable on $\mathbf{R}^p$.

Exercise 7 can be used to construct integrable functions on $\mathbf{R}^p$. For instance, let f be constant on H_1 and on each set $H_n - H_{n-1}$, with values c_1 and c_n, respectively. Let $\sum_{n=1}^{\infty} n^{p-1} |c_n|$ be finite. Then f is absolutely integrable over $\mathbf{R}^p$. Its integral is easily found in series form.

Exercise 7 confirms integrability over $\mathbf{R}^p$ in many examples in which continuity assures integrability on the bounded intervals in $\mathbf{R}^p$. For instance when $f(x) = e^{-|x|}$ with $|x| = [\sum_{i=1}^{p} x_i^2]^{1/2}$ it suffices to take $b_n = e^{-(n-1)}$. As another illustration let $f(x) = 1$ when $|x| \leqslant 1$ and $f(x) = 1/|x|^\alpha$ when $|x| \geqslant 1$. Now $b_n = (n-1)^{-\alpha}$ for $n \geqslant 2$ and $b_1 = 1$ are bounds for f. Then f is integrable when $\alpha > p$.

Comparison is a convenient way to check absolute integrability. We can now improve the test given in Section 3.2 by assuming that f is integrable on a sequence of sets. We need not take care that the members of the sequence be nonoverlapping or expanding—unless we want to use the sequence to evaluate $\int_E f$.

Exercise 8. (Comparison test.) Let $E = \bigcup_{n=1}^\infty E_n$. Let I be a closed interval containing E. Let f be integrable on each E_n. Let the nonnegative function g be integrable on I. Suppose $|f| \leqslant g$ on E. Show that f is absolutely integrable on E.

This is a full-fledged generalization of the comparison test for absolute convergence of series. The integral over E_n corresponds to a finite sum of terms of the series.

Keep in mind that one sequence of sets may be used to establish integrability of f on E and a second sequence may be used to evaluate $\int_E f$.

Earlier it was shown that continuity of f on a closed bounded set E implies integrability of f on E. The next example shows that continuity on an open set along with a second constraint yields integrability.

Example 3. Let E be open in a bounded interval I. Let f be continuous on E and bounded. Show that $\int_E f$ exists.

Solution. Let us take for granted this fact about open sets in $\mathbf{R}^p$: that there is a sequence $(I_n)_{n=1}^\infty$ of non-overlapping closed intervals whose union is E. Then f and $|f|$ are integrable on each I_n. Since I and f are bounded, a constant function can be taken for the majorant g which is integrable over I. Consequently, f is absolutely integrable on E by the comparison test.

It is easy to supply other conditions on f which will imply existence of $\int_E f$ when E is open.

As a final item let's look again at a problem which was considered in Section S1.8 and Section S2.8. Recall that existence of $\lim_{s \to a} \int_s^b f$ is equivalent to existence of $\int_a^b f$. A simple example, given on page 67, showed that existence of $\lim_J \int_J f$ as J expands within I does not guarantee integrability of f on I when I is two-dimensional. The results of this section do enable us to reach further conclusions about this question. The expanding sequence criterion for nonnegative functions implies equivalence of $\lim_J \int_J f$ to $\int_I f$ when f is nonnegative. With just a little effort one can show that $\lim_J \int_J f = \int_I f$ when f is absolutely integrable on each J and $\int_J |f|$ is bounded as J expands.

The distinction between one dimension and higher dimensions arises from the following geometric fact. Let I be a closed interval. Let I' be one of its subintervals such that $I - I'$ is M-null. (The interval I' is obtained by stripping one or more faces from I.) Let $\mathcal{D}$ be a division of I and let $\mathcal{E}$ be the subset of $\mathcal{D}$ consisting of all intervals in $\mathcal{D}$ which are contained in I'. When I is one-dimensional the union of the intervals in $\mathcal{E}$ is an interval. When I has a higher dimension the union of the intervals in $\mathcal{E}$ need not be an interval. See Figure 1 where I' is I without its right-hand edge. The shaded portion is the union of the intervals of $\mathcal{E}$.

When $\int_J f$ exists for each closed subinterval of I' the correct way to formulate a limit criterion for existence of $\int_I f$ is to use the partial division $\mathcal{E}$ mentioned above if absolute integrability of f is not to be invoked. A complete statement of a criterion of this sort is given in Exercise 28 at the end of the chapter.

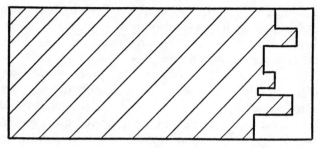

Fig. 1

4.6. Length, area, volume, and measure. In elementary discussions of integrals the notion of area is usually treated as a known and intuitively clear concept. It is used to clarify the motivations behind the definition of an integral of a function defined on an interval of real numbers. Indeed this was done in Section 1.1. But, as we have seen, the area concept itself does not enter into the definition of $\int_a^b f$. It is not difficult to see that any notion of area which has a few natural properties must assign to the region under the graph of f the number $\int_a^b f$ as its area when $f \geqslant 0$. Similarly $\int_a^b (g - f)$ is the area of the region between the graphs of f and g when $f(x) \leqslant g(x)$ for all x in $[a, b]$.

Integrals over real intervals suffice for the calculation of the areas of fairly simple sets. More complex sets do not lend themselves to such an area calculation. Some sets are so complicated that it is not clear on an intuitive level whether it is meaningful to speak of their areas. Consequently, it is necessary to have a comprehensive *definition* of area against which sets can be tested.

When $E \subseteq \mathbf{R}^2$, the area of E is defined to be the integral over E of the constant function whose value is 1. Equivalently, integrate over $\mathbf{R}^2$ the function which is one on E and zero elsewhere.

Notice that there is no circularity in this definition. The only notion of area which goes into it is the area of a bounded interval. The definition is faithful to that primitive notion of area since the integral of 1 over I is $M(I)$ when I is bounded. The natural properties of area, such as additivity, are easily confirmed by specializing the properties of the integral given in Section 4.3. Since the integrable functions form a large class, we have good reason to expect many sets to possess an area according to the definition we have adopted.

Since we have a notion of integral for subsets of $\mathbf{R}^p$ for $p = 1, 2, 3, \ldots$, we can carry out the program proposed for the definition of area in other dimensions, too. In $\mathbf{R}$ we will get an extension of the concept of the length of a bounded interval. In $\mathbf{R}^3$ we get a wider concept of the volume of a three-dimensional set. From now on we will deal with the general situation unless a special instance is made explicit.

In general, when $E \subseteq \mathbf{R}^p$ let χ_E be the *characteristic function* of E, i.e., the function which is 1 on E and 0 elsewhere. Say that E is *integrable* when χ_E is integrable over $\mathbf{R}^p$. Set the *measure* of E, $\mu(E)$, equal to $\int_{\mathbf{R}^p}\chi_E$ in this case.

When I is a bounded interval in $\mathbf{R}^p$ we know that χ_I is integrable and that $\mu(I) = M(I)$. Thus the measure μ is an extension of the interval measure so long as we think only of bounded intervals. When I is unbounded and nondegenerate, its characteristic function is not integrable. Thus we have, up to this point, assigned no meaning to $\mu(I)$. Later we will adopt the convention that $\mu(I) = \infty$. Thus it will not be the case that M and μ agree on nondegenerate unbounded intervals.

Since χ_E is a null function if and only if E is M-null, we conclude that $\mu(E) = 0$ if and only if E is M-null.

From earlier sections of this chapter we also know that

these are integrable sets: any bounded interval, any closed and bounded set, and any set which is open in a bounded interval.

Set operations are the best source of additional integrable sets. We can utilize previous results, especially those on pages 114 and 115, to draw conclusions about the effect of set operations on integrable sets and the properties of the measure function. First we will consider finite operations.

Suppose E, F, E_1, E_2, . . . , E_n are integrable sets. Then:

(i) $0 \leqslant \mu(E)$.

(ii) $E \cup F$ and $E \cap F$ are integrable and

$$\mu(E \cup F) = \mu(E) + \mu(F) - \mu(E \cap F).$$

(iii) $F - E$ *is integrable. When* $E \subseteq F$, $\mu(F - E) = \mu(F) - \mu(E)$.

(iv) $\bigcup_{i=1}^{n} E_i$ and $\bigcap_{i=1}^{n} E_i$ are integrable and

$$\mu\left(\bigcup_{i=1}^{n} E_i \right) \leqslant \sum_{i=1}^{n} \mu(E_i).$$

(v) *If* $E_i \cap E_j$ *is M-null when* $i \neq j$, *then* $\mu(\bigcup_{i=1}^{n} E_i) = \sum_{i=1}^{n} \mu(E_i)$.

Clearly property (v) is finite additivity. We can extend it to countable additivity by specializing the proposition on page 119.

Suppose $(E_n)_{n=1}^{\infty}$ *are integrable sets and* $E_i \cap E_j$ *is M-null when* $i \neq j$. *Then* $\bigcup_{n=1}^{\infty} E_n$ *is integrable if and only if* $\sum_{n=1}^{\infty} \mu(E_n)$ *is finite. Moreover* $\mu(\bigcup_{n=1}^{\infty} E_n) = \sum_{n=1}^{\infty} \mu(E_n)$.

Without the assumption that the sets are nonoverlapping, integrability of the union of a sequence hasn't such a neat formulation. Integrability of a union fails, when it fails, because the union covers too much of $\mathbf{R}^p$—perhaps all of it. Thus some condition must limit the extent of the union. The next exercise offers two different ones.

Exercise 9. Let E_n, $n = 1, 2, 3, \ldots$, be integrable sets. Show that $\bigcap_{n=1}^{\infty} E_n$ is integrable. Show, moreover, that $\bigcup_{n=1}^{\infty} E_n$ is integrable if $\sum_{n=1}^{\infty} \mu(E_n)$ is finite or if $\bigcup_{n=1}^{\infty} E_n$ is a subset of an integrable set.

In summary, the collection of integrable sets is closed under all the standard set operations except for unions of sequences. Yet those unions of integrable sets which fail to be integrable have many useful properties. It is appropriate to define a larger class of sets, the *measurable* sets, which includes them.

Definition. A subset E of $\mathbf{R}^p$ is *measurable* when $E \cap J$ is integrable for every closed, bounded interval J.

Each integrable set is measurable because the intersection of any two integrable sets is integrable. It is clear from the definition of measurable set that each bounded measurable set is integrable. It is customary to set $\mu(E) = \infty$ when E is measurable but not integrable. (As noted above, $\mu(I) = \infty$ but $M(I) = 0$ when I is unbounded and nondegenerate.)

The introduction of ∞ as a value of μ makes possible a satisfyingly simple relation between $\mu(\bigcup_{n=1}^{\infty} E_n)$ and $\sum_{n=1}^{\infty} \mu(E_n)$ provided we agree that the value of the series is ∞ whenever one or more terms equals ∞ or this is a divergent series of nonnegative real numbers. The next exercise sets out the basic facts about operations on measurable sets.

Exercise 10. (a) Suppose E and F are measurable. Show that $F - E$, $\mathbf{R}^p - E$, $E \cap F$, and $E \cup F$ are all measurable. (b) Show that $\bigcap_{n=1}^{\infty} E_n$ and $\bigcup_{n=1}^{\infty} E_n$ are measurable when all E_n are measurable and that $\mu(\bigcup_{n=1}^{\infty} E_n) \leqslant \sum_{n=1}^{\infty} \mu(E_n)$. (c) Show that $\mu(\bigcup_{n=1}^{\infty} E_n) = \sum_{n=1}^{\infty} \mu(E_n)$ when $\mu(E_i \cap E_j) = 0$, $i \neq j$.

The application of the operations in Exercise 10 to sets known to be integrable will yield measurable sets. Since bounded closed and open sets are integrable, all closed sets and all open sets are measurable. Further application of the countable union and intersection will yield other measurable sets which are neither open nor closed. A more definitive examination of the class of measurable sets will be made in Section S8.1.

The measurable sets play a special role in absolute integrability over noninterval sets.

Suppose f is absolutely integrable on an interval I. Then f is absolutely integrable on each measurable subset of I.

The usual reduction brings the question down to a nonnegative integrable function f. Let E be a measurable subset of I and set $f_n = f \wedge (n\chi_E)$. The functions f_n are integrable and f dominates them. They increase to a limit which agrees with f on E and vanishes on $I - E$. Thus $\int_E f$ exists according to the monotone convergence theorem.

In the reverse direction the question is whether existence of $\int_E |f|$ implies measurability of E. Since the zero function is integrable on every set, a distinction must be made between nonzero values and zero values of f.

Suppose $|f|$ is integrable on each bounded subinterval of an interval I. Set $E = \{x \in I : f(x) \neq 0\}$. Then E is measurable. Moreover, when $\int_F |f|$ exists, $E \cap F$ is measurable.

To show E is measurable we need to show $E \cap J$ is integrable when J is a bounded subinterval of I. The integrable functions $(n|f|) \wedge \chi_J$ increase to $\chi_{E \cap J}$ and are dominated by the integrable function χ_J. Thus $\chi_{E \cap J}$ is integrable and E is measurable.

The measurability of $E \cap F$ is a corollary obtained by replacing the values of f by zero outside F.

The measurable sets clearly play a major role in integration over sets which are not intervals. We know how to create many measurable sets by repeated application of the set operations to open and closed sets. Perhaps every subset of $\mathbf{R}^p$ is measurable. No, there are nonmeasurable sets. The examples are not easy to come by, however. The construction of such sets relies on the set-theoretic principle called the axiom of choice.

Those readers who would like to see the details of the construction of a nonmeasurable subset of $[0, 1]$ may consult [9, p. 157] or almost any other work on measure and integration. The properties of measure which have been developed above suffice for an understanding of such a construction. Of course one should substitute the measure μ above for the measure used in the reference.

The existence of a nonmeasurable subset E of $[0, 1]$ clarifies the relation between integrability of f and $|f|$. It has already been shown that $\int_0^1 |f|$ need not exist when $\int_0^1 f$ exists, because offsetting positive and negative values of f may produce a positive function $|f|$ which is too large for existence of the integral. To create an example in which $|f|$ is integrable but f is not we must rely on irregular behavior of f which is rubbed out in the passage to $|f|$. For a nonmeasurable subset E of $[0,1]$, set $f(x) = 1$ when $x \in E$ and $f(x) = -1$ when $x \in [0, 1] - E$. Since $|f(x)| = 1$ on $[0, 1]$, it is true that $\int_0^1 |f|$ exists. But $\chi_E = (f + |f|)/2$. Thus existence of $\int_0^1 f$ implies existence of $\int_0^1 \chi_E$, contrary to the nonmeasurability of E.

4.7. Exercises.

11. Show that the only continuous null function $f : I \to \mathbf{R}^q$ is the one which vanishes everywhere on I.

12. Let $f : E \to \mathbf{R}$ be given. Suppose that there is a gauge γ such that $|fM(\mathcal{S})| < \epsilon$ for all γ-fine partial divisions $\mathcal{S}$ whose tags are in E. Show that E is fM-null.

13. (a) Give an example to show that countable additivity of the integral of f may fail when $|f|$ is not integrable.

(b) Show by example that f need not be integrable on every measurable subset of I when $\int_I f$ exists but $\int_I |f|$ does not.

14. Show that $\sum_{n=1}^{\infty} \int_{E_n} |f|$ cannot be replaced by $\sum_{n=1}^{\infty} |\int_{E_n} f|$ in the statement of the countable additivity of the integral on page 119.

15. Show $\exp(-|x|^2)$ is integrable on $\mathbf{R}^p$.

16. Let $I = [-1, 1] \times \cdots \times [-1, 1]$ in $\mathbf{R}^p$. Let $f(x) = 1/|x|^{\alpha}$, $\alpha > 0$. Find a restriction on α which insures that $\int_I f$ exists.

17. Let $E = \{(x, y) : x \geqslant 1 \text{ and } 0 \leqslant y \leqslant 1/x^r\}$. For what values of r is E an integrable set?

18. Let X be a subset of $\mathbf{R}^p$ with the following property. To each ϵ there corresponds a sequence of open bounded intervals G_n such that $X \subseteq \bigcup_{n=1}^{\infty} G_n$ and $\sum_{n=1}^{\infty} M(G_n) < \epsilon$. Show that X is M-null.

Can you prove the converse too?

19. Let $E \subseteq F \subseteq G$. Let E and G be measurable sets with $\mu(G - E) = 0$. Show that F is measurable and that $\mu(F) = \mu(E) = \mu(G)$.

20. Let E_n be measurable for $n = 1, 2, 3, \ldots$. (a) Suppose $E_n \subseteq E_{n+1}$ for all n. Show that $\mu(\bigcup_{n=1}^{\infty} E_n) = \lim_{n \to \infty} \mu(E_n)$. (b) Suppose $E_{n+1} \subseteq E_n$ for all n and $\mu(E_n) < \infty$ for some n. Show that $\mu(\bigcap_{n=1}^{\infty} E_n) = \lim_{n \to \infty} \mu(E_n)$. (c) Show that $\mu(\bigcap_{n=1}^{\infty} E_n) < \infty$ is possible while $\mu(E_n) = \infty$ for all n.

21. The Cantor set D is a subset of $[0, 1]$ defined as follows. Let D_1 be the set remaining after removal of the middle third $(\frac{1}{3}, \frac{2}{3})$ from $[0, 1]$. Let D_2 be the remaining set after the removal of the middle thirds $(\frac{1}{9}, \frac{2}{9})$ and $(\frac{7}{9}, \frac{8}{9})$ from the intervals of D_1. Continue in this manner. Let $D = \bigcap_{n=1}^{\infty} D_n$. Show that D is an L-null closed subset of $[0, 1]$. Show that D is not a countable set.

22. The Cantor function $F : [0, 1] \to [0, 1]$ is defined in steps as follows. Initially set $F(0) = 0$ and $F(1) = 1$. On the middle third $(\frac{1}{3}, \frac{2}{3})$ let $F(x) = \frac{1}{2}$. Note that this is the average of its values to the left and right. On $(\frac{1}{9}, \frac{2}{9})$ let $F(x) = \frac{1}{4}$ and on $(\frac{7}{9}, \frac{8}{9})$ let $F(x) = \frac{3}{4}$. Continue in this fashion, assigning on a middle third the average of the values on the nearest intervals to the left and right on which F is already defined.

This defines F except on the Cantor set D. Show that F can be defined on D so that it is continuous on $[0, 1]$. Show, moreover, that F is nondecreasing and $F'(x) = 0$ for all $x \notin D$.

Use F to give another proof that D is not countable.

23. Modify the construction of the Cantor set D of Exercise 21 as follows. Let $a_0, a_1, a_2, \ldots$ be positive numbers with $\sum_{n=0}^{\infty} a_n = r$ and $r \leqslant 1$. From the middle of $[0, 1]$ remove an open interval of length a_0. From the middle of each of the remaining two intervals remove an open interval of length $2^{-1} a_1$. At the next stage remove an open interval of length $2^{-2} a_2$ from the middle of each of the remaining four intervals. Continue in this fashion to form a set analogous to the set D of Exercise 21. Does this set have any interior points? What is its measure?

24. Let $[a, b] \subseteq \mathbf{R}$. A function $g : [a, b] \to \mathbf{R}^q$ is said to be *absolutely continuous* provided that to each ϵ there corresponds a δ such that $|\Delta g|(\mathcal{S}) < \epsilon$ for every finite set $\mathcal{S}$ of nonoverlapping intervals satisfying $L(\mathcal{S}) < \delta$.

(a) Suppose f is a bounded integrable function. Set $g(x) = \int_a^x f$, $a \leqslant x \leqslant b$. Show that g is absolutely continuous.

(b) Show that the Cantor function of Exercise 22 is not absolutely continuous.

25. Prove this version of the fundamental theorem: *Let F be absolutely continuous on $[a, b]$. Suppose $F'(x) = f(x)$ except possibly on an L-null subset of $[a, b]$. Then f is integrable and $\int_a^b f = F(b) - F(a)$.*

26. Let E be an open set in $\mathbf{R}^p$. Let f be integrable on each closed interval in E. Give a necessary and sufficient condition for the absolute integrability of f over E.

27. Let I' be a subinterval of the closed interval I which is open in I. Let $f : I' \to \mathbf{R}^q$ be integrable on each closed subinterval of I'. Construct a gauge γ on I' so that $|fM(\mathcal{D}) - \int_J f| < \epsilon$ whenever J is a closed subinterval of I' and $\mathcal{D}$ is a γ-fine division of J. (Compare this with the one-dimensional case in Section S2.8.)

28. Make the same assumptions as in the preceding exercise. Prove that f is integrable on I' if and only if the following limit exists.

Let $\nu(J) = \int_J f$ when $J \subseteq I'$ and $\nu(J) = 0$ when $J \subseteq I$ but $J \not\subseteq I'$. Say that $A = \lim_{\mathcal{D}} \nu(\mathcal{D})$ provided there exists a gauge γ on $I - I'$ such that $|A - \nu(\mathcal{D})| < \epsilon$ whenever $\mathcal{D}$ is a division of I whose subset having tags in $I - I'$ is γ-fine.

29. Let $(f_n)_{n=1}^\infty$ be functions each of which has a primitive on $[a, b]$. On $[0, \infty) \times [a, b]$ set $f(x, y) = f_n(y)$ when $n - 1 \leqslant x < n$. (a) Show that f is integrable on $[n - 1, n] \times [a, b]$ for each positive n. (b) Find conditions under which f is integrable on $[0, \infty) \times [a, b]$.

CHAPTER 5

MEASURABLE FUNCTIONS

From the first example in Chapter 1 it has been clear that a function need not be highly regular in order to be integrable. Yet it cannot be wildly irregular. So far we have relied on two kinds of hypotheses to provide sufficient regularity to insure integrability. One type assumes integrability on certain subsets, say all bounded intervals contained in a given unbounded interval. In the other it is the relation of the function to one or more other functions which supplies the appropriate properties. There are two obvious instances of this. One is the relation of $|f|$ to f when f is integrable and integrability of $|f|$ is in question. Convergence theorems are a second. In each of these instances the function under examination gets its regularity "by inheritance," one might say, from the integrability of other functions.

Since the behavior of Riemann sums provides the criterion for existence or nonexistence of an integral, it has been possible to go very far without identifying the kind of regularity which underlies integration. There are problems which are much easier to solve when it is known just what sort of regularity goes with integrability. The discussion in Section 5.1 goes only as far as identifying the regularity property of *absolutely* integrable functions.

The appropriate concept is measurability of functions. It is expressed in terms of measurable sets. The function which is one on a measurable set and zero elsewhere is the simplest of the measurable functions. The general definition is an analogue of one of the standard formulations of continuity. The definition and illustrative examples occupy Section 5.1.

Section 5.2 makes two main points. One is that absolute integrability implies measurability. The other is that a measurable function which has an integrable dominant is absolutely integrable. This is another comparison test for absolute integrability. Along the way to the second conclusion it is necessary to show that measurable functions can be approximated by measurable functions of a special kind, namely those which take only a finite set of values. The approximation theorem has many other uses, too.

Measurable functions are easy to use. This is largely due to the fact that many operations yield measurable functions when they are applied to measurable functions. Sum, product, lattice operations, and absolute value are among them. And the limit of any sequence of measurable functions is measurable. This is probably the most useful characteristic of measurable functions. Note that no special kind of convergence is needed. Pointwise existence of the limit is enough to transfer measurability of the members of the sequence to the limit function. The justifications of these properties are surprisingly easy. They may be found in the small compass of Section 5.3.

The integrability of products is difficult to get at by use of integrable functions alone. Since the product of measurable functions is measurable, the absolute integrability of a product rests on finding a suitable integrable dominant. Some of the standard results are given in

Section 5.4. These include absolute integrability of the product when one factor is bounded and the other is absolutely integrable and also when the square of the absolute value of each factor is integrable.

The definition of measurability of functions utilizes measurable sets which hark back to integrability, too. It is appropriate to ask how measurability of a function can be characterized in an integration-free manner. The answer is that measurable functions are those which are limits of sequences of step functions. Section S5.5 demonstrates this. Along the way there are interesting facts about the approximation of absolutely integrable functions by step functions. These propositions are not among the basic tools for the user of integration theory. Hence this section has been marked to be skipped on first reading.

5.1. Measurable functions. Continuity is the most familiar indicator of the regularity of a function. One of the standard characterizations of continuity lends itself to a generalization which is natural to the discussion of integration. This generalization is the concept of measurability of functions.

Let $f : I \to \mathbf{R}^q$ be given. Let $E \subseteq \mathbf{R}^q$. The *inverse image* of E by f, denoted $f^{-1}(E)$, is defined by

$$f^{-1}(E) = \{ x \in I : f(x) \in E \}.$$

An examination of the usual ϵ, δ-definition of continuity shows that f is continuous on I if and only if $f^{-1}(G)$ is open in I for every open interval G in $\mathbf{R}^q$.

Here is the generalization we want. A function $f : I \to \mathbf{R}^q$ is *measurable* provided $f^{-1}(G)$ is a measurable subset of I for every open interval G in $\mathbf{R}^q$.

Since an open interval in $\mathbf{R}^q$ is a Cartesian product of open real intervals, a straightforward argument shows that the measurability of f is equivalent to the measurability of all its real-valued components when $f = (f_1, f_2, \ldots, f_q)$. Thus what we learn about measurable real-valued functions can be translated into statements about vector functions if the need arises.

The definition of measurability actually requires the examination of more inverse images than are necessary for the purpose. The reduction to a smaller class of intervals is easier to state for real-valued functions.

Exercise 1. Let $f : I \to \mathbf{R}$ be given. Show that the following four conditions are equivalent: $f^{-1}(G)$ is measurable for all intervals G of the form (i) (a, ∞), (ii) $(-\infty, a]$, (iii) $(-\infty, a)$, (iv) $[a, \infty)$. Also show that each of these conditions is equivalent to measurability of f.

It is clear from Exercise 1 that $f^{-1}(G)$ is measurable for every interval G, open or not, when f is measurable.

What functions are measurable? Since sets which are open in I are measurable, it is evident that every continuous function is measurable. Even less regularity than continuity everywhere implies measurability. For instance, let $f(x) = 1$ when x is irrational and $f(x) = 0$ when x is rational. Then $f^{-1}(G)$ is one of four sets—the empty set, the rationals, the irrationals, or all of the reals. All of them are measurable. Thus f is measurable. But f is not continuous anywhere. By looking at restrictions of f one can identify some continuity, however. The next example generalizes the connection we see here between continuity and measurability.

Example 1. Let $f : I \to \mathbf{R}^q$ have a continuous restriction to a set A such that $I - A$ is M-null. Show that f is measurable.

Solution. Let G be an open interval and consider $f^{-1}(G)$. It suffices to show that $A \cap f^{-1}(G)$ and $(I - A) \cap f^{-1}(G)$ are measurable. The latter is measurable since it is M-null. Consider the former and let $y \in A \cap f^{-1}(G)$. Let's apply the continuity of the restriction of f. There is an open interval J containing y such that $f(x) \in G$ for all x in $A \cap J$. Let H be the union of such open intervals obtained by varying y over $A \cap f^{-1}(G)$. Then $A \cap f^{-1}(G) = A \cap H$. Now A is measurable because it is the complement of the M-null set $I - A$. Clearly H is open, hence measurable. Thus $A \cap H$ is measurable.

When A is an open set, the continuity of the restriction of f to A is equivalent to continuity of f itself at each point of A. This is a contrast with the instance above in which A is the set of irrationals. There are cases of practical importance in which A can be chosen to be an open set, however. The next example provides an illustration.

Example 2. Let E be the plane region bounded by the graphs of two continuous functions. Let $f : E \to \mathbf{R}$ be continuous. Let $g(x) = f(x)$ when $x \in E$ and $g(x) = 0$ otherwise. Show that g is measurable.

Solution. Let B be the boundary of E and let $A = \mathbf{R}^2 - B$. Since B consists of vertical line segments and two continuous function graphs, B is M-null and A is open. Clearly g is continuous at interior points of E because f is continuous. At points exterior to E, g is continuous because it is constant outside E. Thus g is continuous on A.

5.2. Measurability and absolute integrability. Now it is time to explore the connection between measurability and integrability. Actually it is *absolute* integrability that gives the connection we want.

Let $f : I \to \mathbf{R}^q$ be absolutely integrable on each bounded subinterval of I. Then f is measurable.

We may suppose f is real-valued. Exercise 1 allows us to confine our attention to inverse images of intervals (a, ∞). Let χ_I be the characteristic function of I, i.e., the function which is one on I and zero elsewhere. Set $g = 0 \vee (f - a\chi_I)$. Then $g(x) \neq 0$ if and only if $x \in f^{-1}((a, \infty))$. Since f and χ_I are absolutely integrable on each bounded interval, g enjoys that property too. Consequently, the set where g is nonzero is measurable. (See p. 126.) Thus f is measurable.

This proposition informs us that absolutely integrable functions are to be found among the measurable functions. But not all measurable functions are absolutely integrable on bounded intervals. Example 1 provides examples to the contrary, for instance $f(0) = 0$ and $f(x) = 1/x$ when $0 < x \leqslant 1$. This function fails to be integrable not from lack of regularity but from the presence of too many large values. The same is true for nonintegrable measurable functions in general, as we shall see presently.

What we aim toward is yet another version of the comparison test for absolute integrability.

COMPARISON TEST. *If f is measurable, g is integrable, and $|f| \leqslant g$, then f is absolutely integrable.*

This may be regarded as the ultimate version since the regularity which is being imposed on f, i.e., measurability, is as little as it could possibly possess.

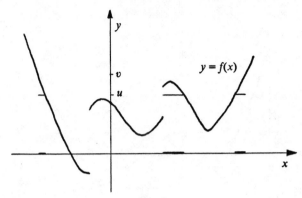

FIG. 1

To reach this goal we must be able to approximate measurable functions with absolutely integrable functions in order to apply a convergence theorem. A bit of exploratory work will reveal the special kind of functions we should use.

Consider a measurable function $f : I \to \mathbf{R}$. For the moment, make two restrictions. Suppose I is a bounded interval and f is a bounded function with $-M < f(x) < M$ for all x in I. The set $E = \{x \in I : u \leqslant f(x) < v\}$ is a measurable set, since it is $f^{-1}([u, v))$. Think of u and v as close together. Then $u\chi_E$ is a good approximation to f on E. (See Fig. 1.) Since I is bounded χ_E and $u\chi_E$ are integrable. To approximate f on all of I it is necessary to decompose I into a finite collection of sets of this sort. This will be the next step.

Form a division of $[-M, M]$ into n closed intervals $[y_{k-1}, y_k]$ of common length $2M/n$. The half-open intervals $[y_{k-1}, y_k)$ cover the interval $(-M, M)$ in which all values of f lie. The corresponding sets $E_k = f^{-1}([y_{k-1}, y_k))$ cover I and no two have common points.

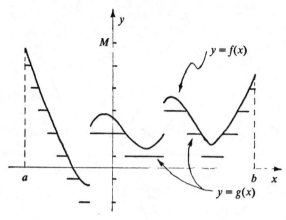

FIG. 2

Set $g = \sum_{k=1}^{n} y_{k-1} \chi_{E_k}$. Then $g(x) = y_{k-1}$ when $x \in E_k$. Consequently $|f(x) - g(x)| \le 2M/n$ for all x in I. (See Fig. 2.) Moreover g is absolutely integrable since each set E_k is integrable.

Note that the approximation is equally good if some other number in $[y_{k-1}, y_k]$ is used in place of y_{k-1} in forming g. Suitable choices will yield a function g such that $g \le f, f \le g$, or $|g| \le |f|$.

A function is a *simple* function when it is measurable and takes a finite set of values. It is immediate that each value is assumed on a measurable set. When each nonzero value is assumed on an integrable set the simple function is absolutely integrable. Thus the function g above is an absolutely integrable simple function.

The procedure above can be used to generate a sequence of integrable simple functions $(f_n)_{n=1}^{\infty}$ which converge to f. With a little care the sequence can be chosen so that it increases, decreases, or has increasing absolute values. These properties are achieved by using a sequence of divisions of $[-M, M]$, each of which is a refinement of its predecessor. In the present case of a bounded function f on a bounded interval the convergence of the sequence is uniform. More generally, when the interval I or the function f is not bounded, it is necessary to give up uniformity of convergence. Enough other properties can still be achieved to make the sequence amenable to a convergence theorem.

Let $f : I \to \mathbf{R}^q$ be measurable. There exists a sequence $(f_n)_{n=1}^\infty$ of absolutely integrable simple functions such that $\lim_{n\to\infty} f_n(x) = f(x)$ at every finite point x in I. Let $(I_n)_{n=1}^\infty$ be an expanding sequence of bounded closed intervals whose union is $I \cap \mathbf{R}^p$. The function f_n can be defined so that f_n vanishes outside I_n, $|f_n| \le |f|$, and $|f_n| \le |f_{n+1}|$ for all n. Each f_n can be made nonnegative when f is nonnegative.

Exercise 2. Construct a sequence of functions with these properties.

The dominated convergence theorem brings us immediately to our goal of a proof of the comparison test for absolute integrability.

5.3. Operations on measurable functions. Measurable functions are easy to use largely because so many operations on measurable functions yield measurable functions. The following general result has many important special instances.

The composite of a measurable function followed by a continuous function is measurable.

More precisely, let H be an open set in $\mathbf{R}^q$. Let

$f: I \rightarrow H$ be measurable and $T: H \rightarrow \mathbf{R}^t$ be continuous. Then $T \circ f$ is measurable.

Let G be an open interval in $\mathbf{R}^t$. We must show that $(T \circ f)^{-1}(G)$ is measurable. It is easy to see that $(T \circ f)^{-1}(G) = f^{-1}(T^{-1}(G))$. Now $T^{-1}(G)$ is an open set since T is continuous on the open set H. Any open set is the union of a sequence of intervals. Thus $T^{-1}(G) = \bigcup_{n=1}^{\infty} I_n$. A quick check confirms the equation

$$f^{-1}\left(\bigcup_{n=1}^{\infty} I_n \right) = \bigcup_{n=1}^{\infty} f^{-1}(I_n).$$

Each set $f^{-1}(I_n)$ is measurable. A countable union of measurable sets is also measurable. Thus $(T \circ f)^{-1}(G)$ is measurable, as required.

This one assertion about composites yields several conclusions about algebraic and lattice operations on measurable functions.

Let f and g be measurable. Let c be constant. Each of the following is also measurable: $cf, f + c, |f|^t$ with $t > 0$, $f + g, fg, f/g$ when g has no zero values, $f \vee g$, and $f \wedge g$.

The proofs all result from the choice of an appropriate T in each case. For instance, to deal with $f + g$ let $T(u, v) = u + v$ on $\mathbf{R}^q \times \mathbf{R}^q$ and consider $T \circ (f, g)$. Of course it must be noted that T is continuous and (f, g) is measurable. The other cases are handled similarly.

When f is real-valued f^+ and f^- are measurable since $f^+ = f \vee 0$ and $f^- = (-f) \vee 0$.

A function which is zero almost everywhere is a measurable function. Thus a function which is equal almost everywhere to a measurable function is measurable, since it is the sum of two measurable functions.

It is most important that *the limit of a sequence of measurable functions is measurable*. This is an easy consequence of the equation given in the next exercise.

Exercise 3. Suppose $\lim_{n\to\infty} f_n(x) = f(x)$ for all x in I. Suppose the functions are real-valued. Let $G = (a, \infty)$ and $G_n = (a + 1/n, \infty)$. Show that

$$f^{-1}(G) = \bigcup_{n=1}^{\infty} \bigcup_{m=1}^{\infty} \bigcap_{k=m}^{\infty} f_k^{-1}(G_n).$$

It is enough to assume convergence almost everywhere in order to deduce measurability of the limit. This is because the functions can be modified in an M-null set to achieve convergence everywhere without losing measurability.

Example 3. Suppose g is the derivative of f almost everywhere. Show that f and g are measurable.

Solution. f is continuous wherever it has a derivative, hence almost everywhere. Thus, by Example 1, f is measurable. Now let $g_n(x) = n(f(x + 1/n) - f(x))$. Then g_n is continuous almost everywhere and $\lim_{n\to\infty} g_n(x) = f'(x) = g(x)$ almost everywhere. Consequently g is the limit of measurable functions. Thus g is measurable.

5.4. Integrability of products. The product of measurable functions is measurable. When the factors have properties which yield an integrable dominant for the product, the comparison test tells us that the product is integrable.

Suppose f is absolutely integrable and g is bounded and measurable. Then fg is absolutely integrable.

Let M be a bound on g. Then $|f| \leqslant M|f|$. Thus $M|f|$ is the desired integrable dominant.

Example 4. (a) Let $f(x) = g(x) = x^{-1/2}$ when $x > 0$. Then f is absolutely integrable on $[0, 1]$ but fg is not. Hence boundedness of g cannot be left out of the above assertion.

(b) Let f be the function of Example 2 on page 79. Remember that $|f|$ is not integrable. Let $g(x) = 1$ on those intervals where $f(x) \geqslant 0$ and $g(x) = -1$ on the intervals where $f(x) < 0$. Then g is bounded and obviously measurable. Since $fg = |f|$, the product fg is not integrable. Consequently, absolute integrability of f cannot be weakened to integrability in the above proposition.

An integrable dominant of fg exists when certain powers of $|f|$ and $|g|$ are integrable.

Suppose f and g are measurable and $|f|^s$ and $|g|^t$ are integrable where $s > 1$, $t > 1$, and $1/s + 1/t = 1$. Then fg is absolutely integrable.

A simple inequality is all we need. Suppose $a \geqslant 0$ and $b \geqslant 0$. Either $a \leqslant b^{t-1}$ or $a > b^{t-1}$. In the first case $ab \leqslant b^t \leqslant a^s + b^t$. In the second case $a^{s-1} > b^{(t-1)(s-1)}$. However, a few algebraic steps show that $(t-1)(s-1) = 1$ when $1/s + 1/t = 1$. Thus $b < a^{s-1}$ and $ab < a^s \leqslant a^s + b^t$. In summary, $ab \leqslant a^s + b^t$ when $a \geqslant 0$ and $b \geqslant 0$.

Now $|fg| = |f| \, |g| \leqslant |f|^s + |g|^t$. We have the integrable dominant we need in order to conclude that fg is absolutely integrable.

The inequality $|fg| \leqslant |f|^s + |g|^t$ also yields an estimate on the integral of $|fg|$, namely,

$$\int_I |fg| \leqslant 2 \left(\int_I |f|^s \right)^{1/s} \left(\int_I |g|^t \right)^{1/t}.$$

However, there is a sharper inequality on $|fg|$ which yields the sharper integral estimate in which the factor 2 is replaced by 1.

Exercise 4 (Hölder's inequality). Take for granted the inequality $|ab| \leqslant |a|^s/s + |b|^t/t$ when $s > 1$, $t > 1$, and $1/s + 1/t = 1$. Show that

$$\int_I |fg| \leqslant \left(\int_I |f|^s \right)^{1/s} \left(\int_I |g|^t \right)^{1/t}$$

when f and g are measurable and $|f|^s$ and $|g|^t$ are integrable.

A justification for the inequality taken for granted in Exercise 4 is outlined in Exercises 7 and 8 at the end of the chapter.

S5.5. Approximation by step functions. Step functions are the simplest of the simple functions. Since step functions are defined independently of the concept of integral, it is of considerable interest to know to what extent measurable and integrable functions can be approximated by step functions. The development of this topic begins with a proposition which is not explicitly tied to the integral concept.

COVERING LEMMA. *Let I be a closed interval and let E be a nonempty subset of $I \cap \mathbf{R}^p$. Let γ be a gauge defined (at least) on E. There exists a sequence $(z_n J_n)_{n=1}^{\infty}$ of tagged nonoverlapping bounded closed intervals such that $z_n \in E$ and $J_n \subseteq \gamma(z_n)$ for all n and $E \subseteq \bigcup_{n=1}^{\infty} J_n \subseteq I$.*

Let $I_n = [-n, n] \times [-n, n] \times \cdots \times [-n, n]$. Let $\mathscr{D}_n$ be the division of I_n obtained by dividing each factor $[-n, n]$

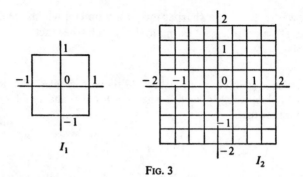

FIG. 3

into $n2^n$ intervals of common length $1/2^{n-1}$. (Figure 3 shows $\mathfrak{D}_1$ and $\mathfrak{D}_2$ in two-dimensional space.) The intervals of $\mathfrak{D}_{n+1}$ which are in I_n are obtained by halving the edges of the intervals of $\mathfrak{D}_n$. Consequently, given J in $\mathfrak{D}_n$ and K in $\mathfrak{D}_{n+1}$, either $K \subseteq J$ or J and K are nonoverlapping.

Let $\mathcal{E}_1$ be the collection of all intervals J in $\mathfrak{D}_1$ such that $J \subseteq \gamma(z)$ for some z in $E \cap J$. Form $\mathcal{E}_2, \mathcal{E}_3, \ldots$ recursively from $\mathfrak{D}_2, \mathfrak{D}_3, \ldots$ as follows. Having formed $\mathcal{E}_{n-1}$, adjoin to it all those intervals J in $\mathfrak{D}_n$ such that (i) J is not contained in an interval of $\mathcal{E}_{n-1}$ and (ii) $J \subseteq \gamma(z)$ for some z in $E \cap J$. The result is $\mathcal{E}_n$.

Let $\mathcal{E} = \bigcup_{n=1}^{\infty} \mathcal{E}_n$. Then $\mathcal{E}$ is a countable (possibly finite) collection of nonoverlapping closed intervals. Let F be the union of all the members of $\mathcal{E}$. To see that $E \subseteq F$, let $z \in E$. Since the intervals I_n expand and have $\mathbf{R}^p$ as their union, there is an integer n_1 such that $z \in I_n$ when $n \geqslant n_1$. When $n \geqslant n_1$, there is at least one interval of $\mathfrak{D}_n$ to which z belongs. Since $\gamma(z)$ is an open interval, any interval containing z with sufficiently short edges is contained in $\gamma(z)$. Thus there is an integer n_2 such that each interval of $\mathfrak{D}_n$ containing z is a subset of $\gamma(z)$ when $n \geqslant n_2$. Fix an n larger than n_1 and n_2 and an interval J in $\mathfrak{D}_n$ which contains z. Then J satisfies (ii) in the definition of $\mathcal{E}_n$. Either $J \in \mathcal{E}_n$ or $J \subseteq K$ for some K in $\mathcal{E}_{n-1}$. In either case $z \in F$.

Now intersect each interval in $\mathcal{E}$ with I. No points of E are lost since $E \subseteq I$. Arrange the resulting intervals as a sequence $J_1, J_2, J_3, \ldots$. Each J_n can be assigned a tag z_n so that $z_n \in E$ and $J_n \subseteq \gamma(z_n)$. This sequence of tagged intervals has all the properties claimed in the lemma.

The covering lemma provides a link between integrable sets and sequences of intervals.

Let E be an integrable subset of $\mathbf{R}^p$. There exists a sequence of closed, bounded, nonoverlapping intervals $(J_n)_{n=1}^{\infty}$ such that $E \subseteq \bigcup_{n=1}^{\infty} J_n$ and $\sum_{n=1}^{\infty} \mu(J_n) \leqslant \mu(E) + \epsilon$.

Let $I = \overline{\mathbf{R}}^p$ and fix γ so that $|\int_I \chi_E - \chi_E M(\mathcal{D})| < \epsilon$ when $\mathcal{D}$ is a γ-fine division of I. Obtain a sequence $(z_n J_n)_{n=1}^{\infty}$ from the covering lemma. Apply Henstock's lemma to the collection $\{z_1 J_1, z_2 J_2, \ldots, z_n J_n\}$. Then $|\sum_{i=1}^{n} [\chi_E M(z_i J_i) - \int_{J_i} \chi_E]| \leqslant \epsilon$. But $\chi_E M(z_i J_i) = \mu(J_i)$ and $\sum_{i=1}^{n} \int_{J_i} \chi_E \leqslant \int_I \chi_E$. Thus $\sum_{k=1}^{n} \mu(J_k) \leqslant \int_I \chi_E + \epsilon$ for all n. The stated conclusion follows by taking the limit on n.

The proposition just proved is remarkably useful. Let us put it to work.

Let E be an integrable set. There exists a set K which is a finite union of nonoverlapping closed intervals such that $\int_I |\chi_E - \chi_K| < \epsilon$.

Using the preceding result with $\epsilon/2$ in place of ϵ, let $F = \bigcup_{i=1}^{\infty} J_i$. Fix n so that $\mu(F) - \epsilon/2 < \sum_{i=1}^{n} \mu(J_i)$ and let $K = \bigcup_{i=1}^{n} J_i$. Now $\int_I |\chi_E - \chi_K| \leqslant \int_I |\chi_E - \chi_F| + \int_I |\chi_F - \chi_K|$. But $\chi_E \leqslant \chi_F$ and $\chi_K \leqslant \chi_F$. Thus $\int_I |\chi_E - \chi_F|$ reduces to $\mu(F) - \mu(E)$. We have chosen F so that $\mu(F) - \mu(E)$

$\leqslant \epsilon/2$. Similarly $\int_I |\chi_F - \chi_K| < \epsilon/2$. The conclusion follows.

Since $\chi_K = \sum_{i=1}^n \chi_{J_i}$ almost everywhere, the preceding result asserts that the characteristic function of an integrable set may be approximated closely by a step function. Two steps carry this process to its natural conclusion, which we state now.

When f is an absolutely integrable function there is a step function g such that $\int_I |f - g| < \epsilon$.

Summation completes the intermediate stage in which absolutely integrable simple functions are to be approximated. When f is absolutely integrable there are simple functions f_n such that $|f_n| \leqslant |f|$ and $\lim_{n \to \infty} f_n = f$. Then $|f - f_n|$ is dominated by $2|f|$ and has a zero limit. Thus we can choose n, using dominated convergence, so that $\int_I |f - f_n| < \epsilon/2$. On approximating f_n by a step function g so that $\int_I |f_n - g| < \epsilon/2$ we get the conclusion $\int_I |f - g| < \epsilon$.

Since $|\int_I f - \int_I g| \leqslant \int_I |f - g|$, the assertion above is stronger than a claim that $|\int_I f - \int_I g| < \epsilon$. The stronger claim is what we need to reach our objective concerning measurable functions.

A function f is measurable if and only if there is a sequence of step functions which converges to f almost everywhere.

Since step functions are measurable, the "if" implication follows from the fact that convergence of measurable functions almost everywhere yields a measurable limit function. For the converse, take a sequence of absolutely integrable simple functions f_n which converges to f. Select a step function g_n so that $\int_I |f_n - g_n| < 1/2^n$. Then

$\sum_{n=1}^{\infty} \int_I |f_n - g_n| < \infty$; consequently $\sum_{n=1}^{\infty} |f_n - g_n|$ converges almost everywhere. Then $\lim_{n\to\infty} |f_n - g_n| = 0$ almost everywhere. We conclude that $\lim_{n\to\infty} g_n = f$ almost everywhere.

5.6. Exercises.

5. Show that an absolutely integrable simple function takes each nonzero value on an integrable set.

6. Produce a simple function which is integrable but not absolutely integrable.

7. (Young's inequality.) Let $f : I \to J$ and $g : J \to I$ be strictly increasing continuous functions which are mutually inverse. Let $a \in I$ and $b = f(a)$. Let $u \in I$ and $v \in J$. Show that

$$uv - ab \leqslant \int_a^u f + \int_b^v g$$

by doing the following:

(a) Draw figures which depict the terms in Young's inequality as areas when (i) $v = f(u)$, (ii) $v < f(u)$, and (iii) $f(u) < v$.

(b) Prove that equality holds in case (i).

(c) Use case (i) to prove inequality in the other two cases.

8. Suppose that $s > 1$, $t > 1$, and $1/s + 1/t = 1$. Deduce the inequality $|ab| \leqslant |a|^s/s + |b|^t/t$ from Young's inequality.

9. Let $(f_n)_{n=1}^{\infty}$ be measurable real-valued functions. Suppose $g(x) = \text{lub}_n f_n(x)$ is finite for all x. Show that g is measurable.

10. Let E be measurable in $\mathbf{R}^p$ and F be measurable in $\mathbf{R}^q$. Show that $E \times F$ is measurable in $\mathbf{R}^{p+q}$.

11. Let $f : \mathbf{R} \to \mathbf{R}$ be measurable. Let $g : \mathbf{R}^p \to \mathbf{R}$ be defined by $g(x_1, x_2, \ldots, x_p) = f(x_1)$. Show that g is measurable.

12. Let E be open in I. Show from the covering lemma that E can be expressed as the union of a sequence of nonoverlapping closed bounded intervals.

CHAPTER 6

MULTIPLE AND ITERATED
INTEGRALS

Two complementary aspects of integration over multi-dimensional intervals are treated in this chapter. First, in Section 6.1, it is assumed that $\int_I f$ exists. Conclusions are asserted concerning the existence of iterated integrals and their equality to $\int_I f$. This is Fubini's theorem. The proof is postponed to Section S6.4. Section 6.2 attacks the complementary question of what can be said about existence of $\int_I f$ from the known properties of an iterated integral. Some necessary conditions are developed which permit the conclusion that $\int_I f$ does not exist. Other conditions are found which allow the conclusion that $\int_I f$ exists to flow from properties of an iterated integral. The behavior of the iterated integral as a function of intervals is of particular interest in these necessary conditions and sufficient conditions for integrability.

Fubini's theorem as stated in Section 6.1 and proved in Section S6.4 does *not* assume absolute integrability. Under the more restrictive hypothesis of absolute integrability an entirely different kind of proof can be given. Such a proof builds up from characteristic functions through simple functions to absolutely integrable functions. It uses convergence theorems in its last stage. The details of this approach are left as Exercise 11 at the end of the chapter.

The proof given in Section S6.4 uses as its main ideas the definition of the integral and the Cauchy criterion. It is simple in concept but requires many details.

The sufficient conditions in Section 6.2 are of two types. One of them, called Tonelli's theorem, is similar to a comparison test. It implies absolute integrability. The others use an iterated integral in a role rather like that of the primitive of a function of one variable in the fundamental theorem. They are not restricted to situations in which absolute integrability is the conclusion.

Section S6.3 is devoted to some technical matters concerning intervals, gauges, and divisions. These are necessary preliminaries to the proof of Fubini's theorem in Section S6.4. They also lend themselves to an induction proof of the compatibility theorem, that is, the assertion that there is a γ-fine division for any gauge γ. (Only the one-dimensional interval was treated in the proof in Section S1.8.)

The material in Section S6.5 concerning double series is not used in the rest of the book. Its purpose is to point out how the generalized Riemann integral can be used as the central concept in double series.

6.1. Fubini's theorem. Fubini's theorem asserts the equality of a multiple integral to the corresponding iterated integrals. When I is the two-dimensional interval $[a, b] \times [c, d]$ there are two equalities. They are $\int_I f = \int_a^b \int_c^d f(x, y)\, dy\, dx$ and $\int_I f = \int_c^d \int_a^b f(x, y)\, dx\, dy$. When $I \subseteq \mathbf{R}^p$ with $p > 2$ there are r-fold iterated integrals for all integers r between 2 and p, inclusive. Fubini's theorem for 2-fold integrals is the key result since repeated 2-fold integration yields r-fold integrals with $r > 2$. Now we shall adopt suitable notation for the statement of Fubini's theorem for 2-fold integrals in $\overline{\mathbf{R}}^p$.

To perform a 2-fold integration the first step is to express $\overline{\mathbf{R}}^p$ as a Cartesian product of two subspaces. Let t and u be positive integers whose sum is p. Select t factors $\overline{\mathbf{R}}$ from $\overline{\mathbf{R}}^p$ and form them into a space $\overline{\mathbf{R}}^t$. (These need not be the first t factors.) Form $\overline{\mathbf{R}}^u$ with the remaining u factors. Now $\overline{\mathbf{R}}^p$ can be identified with $\overline{\mathbf{R}}^t \times \overline{\mathbf{R}}^u$ as follows. Each z in $\overline{\mathbf{R}}^p$ corresponds uniquely to a point (x, y) in $\overline{\mathbf{R}}^t \times \overline{\mathbf{R}}^u$. For instance, when $p = 3$ one choice is to use the first and third factors for $\overline{\mathbf{R}}^t$ and the second for $\overline{\mathbf{R}}^u$. Thus, each point (z_1, z_2, z_3) in $\overline{\mathbf{R}}^3$ corresponds to $((z_1, z_3), z_2)$ in $\overline{\mathbf{R}}^2 \times \overline{\mathbf{R}}$. Note that it is essential to specify not only which factors of $\overline{\mathbf{R}}^p$ go into each of $\overline{\mathbf{R}}^t$ and $\overline{\mathbf{R}}^u$ but also the order in which they occur when there are two or more.

Once the factors and their orders are specified, each interval I in $\overline{\mathbf{R}}^p$ factors in just one way into $G \times H$ where G and H are intervals in $\overline{\mathbf{R}}^t$ and $\overline{\mathbf{R}}^u$. A function $f : I \to \mathbf{R}^q$ can be reinterpreted as a function on $G \times H$ simply by identifying $f(x, y)$ with $f(z)$ when z corresponds to (x, y). The 2-fold iterated integrals are $\int_G \int_H f(x, y) \, dy \, dx$ and $\int_H \int_G f(x, y) \, dx \, dy$. For the purpose of stating and proving Fubini's theorem there is no need to deal with both of these 2-fold integrals. We will use $\int_H \int_G f(x, y) \, dx \, dy$.

(A change of notation arranges matters so that the first integration can always be taken over the first subspace. For example, in the case $p = 3$ we can as well take $x = z_2$ and $y = (z_1, z_3)$ as the reverse.)

Since one or both of x and y may be vector-valued, a comment about the significance of dx and dy is in order. The only role of dx and dy is to show what function is being integrated. In $\int_G f(x, y) \, dx$ it is the function $x \to f(x, y)$, for fixed y, which is being integrated over G. The result is a value of the function $y \to \int_G f(x, y) \, dx$. This is the function which dy signals in $\int_H \int_G f(x, y) \, dx \, dy$.

When there is a need for a symbol for $x \to f(x, y)$, the most convenient choice is $f(\cdot, y)$. Similarly, $\int_G f(x, \cdot)\, dx$ can serve as symbol for the function $y \to \int_G f(x, y)\, dx$. There are similar symbols for the functions which appear in the reverse order of integration. Our agreed notational bias toward $\int_H \int_G f(x, y)\, dx\, dy$ makes the others less needful.

Now it is appropriate to turn from the notational aspects of Fubini's theorem to the precise statement of the conclusions.

FUBINI'S THEOREM.   *Let $I \subseteq \overline{\mathbf{R}}^p$ with $p \geqslant 2$. Suppose $\int_I f$ exists. Let $\overline{\mathbf{R}}^p$ be expressed as $\overline{\mathbf{R}}^t \times \overline{\mathbf{R}}^u$ and let $I = G \times H$. Then $f(\cdot, y)$ is integrable over G almost everywhere on H. Moreover $\int_G f(x, \cdot)\, dx$ is integrable over H and*

$$\int_I f = \int_H \int_G f(x, y)\, dx\, dy.$$

It is to be understood that when $\int_G f(x, y)\, dx$ is not meaningful for all y in H some extension of $\int_G f(x, \cdot)\, dx$ to all of H is to be supplied. Since the set on which such a failure of integrability occurs is M-null in $\overline{\mathbf{R}}^u$, it does not matter how the function is defined at such exceptional points in H.

The proof of Fubini's theorem is straightforward in conception. It contains a considerable number of details. It is just as well to defer it until later and to concentrate first on the significance of the theorem. (See Section S6.4 for the proof.)

When $p \geqslant 3$, at least one of the stages of the iterated integral is subject to another application of Fubini's theorem since $t \geqslant 2$ or $u \geqslant 2$ or both. Repeated application of Fubini's theorem will serve to reduce $\int_I f$ to any r-fold integral we care to specify. An r-fold integral, $2 \leqslant r \leqslant p$, results from the ordered partition of $\overline{\mathbf{R}}^p$ into r

subspaces. This is most easily described by setting forth what happens to a typical point $(z_1, z_2, \ldots, z_p)$ in $\overline{\mathbf{R}}^p$. For instance, in $\overline{\mathbf{R}}^5$, let $x_1 = (z_1, z_3)$, $x_2 = (z_2, z_5)$, and $x_3 = z_4$. Two applications of Fubini's theorem yield this equation:

$$\int_I f = \int_{I_3} \int_{I_2} \int_{I_1} f(x_1, x_2, x_3) \, dx_1 \, dx_2 \, dx_3.$$

A p-fold integration in $\overline{\mathbf{R}}^p$ consists entirely of integrals over intervals in $\overline{\mathbf{R}}$. Thus the fundamental theorem may be applicable in each step of the iterated integration. Here we have one of the most powerful tools for the evaluation of multiple integrals. Note that this procedure presupposes knowing that $\int_I f$ exists. (In Section 6.2 we will explore the reverse question. That is, we will look for properties of the iterated integral which imply existence of $\int_I f$.)

Even though it is stated only for intervals, Fubini's theorem is applicable to integrals over noninterval sets since $\int_E f$ is defined as $\int_I g$ where g agrees with f on E and vanishes elsewhere on an interval I which contains E. The conclusions of Fubini's theorem tell us something about the structure of measurable and integrable sets. The three-dimensional case provides a manageable illustration of this point.

Suppose f is a strictly positive integrable function on $\mathbf{R}^3$. Let E be a measurable subset of $\mathbf{R}^3$. Consider first the iterated integral which results from expressing $\mathbf{R}^3$ as $\mathbf{R} \times \mathbf{R}^2$. Thus $\int_E f = \int_{\mathbf{R}^2} \int_{\mathbf{R}} g(x, y) \, dx \, dy$ where g agrees with f on E and vanishes outside E. In the inner integral $\int_{\mathbf{R}} g(x, y) \, dx$ the function g is positive on the set E_y where $E_y = \{(x, y) \in E : y \text{ is fixed}\}$. According to Fubini's theorem $\int_{\mathbf{R}} g(x, y) \, dx$ exists for almost all y in $\mathbf{R}^2$. Accordingly E_y, or more accurately its projection into $\mathbf{R}$, is a measurable subset of $\mathbf{R}$ for almost all y in $\mathbf{R}^2$. (See Fig. 1(a).)

The reverse arrangement in which $\mathbf{R}^3$ is split as $\mathbf{R}^2 \times \mathbf{R}$

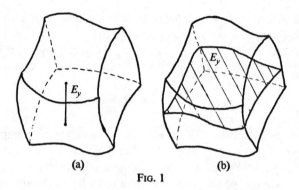

(a)   (b)

FIG. 1

results in E_y being a section of E by a plane parallel to a coordinate plane. In this situation Fubini's theorem assures us that E_y is a measurable two-dimensional set for almost all y in $\mathbf{R}$. (See Fig. 1(b).)

When E is an integrable set, a similar analysis of $\int_{\mathbf{R}^3} \chi_E$ shows that, in the split of $\mathbf{R}^3$ as $\mathbf{R} \times \mathbf{R}^2$, the sets E_y have finite length for almost all y in $\mathbf{R}^2$ and $\mu(E)$ is the integral of the corresponding length function over $\mathbf{R}^2$. In the reverse case, the plane sections of E have finite area for almost all y in $\mathbf{R}$ and $\mu(E)$ is the integral of the area function.

Similar considerations apply in $\mathbf{R}^p$ for other values of p.

6.2. Determining integrability from iterated integrals. In the application of Fubini's theorem, the iterated integral is a means for the evaluation of an integral which is known to exist. It is also possible to prove or disprove the integrability of a function by examining its iterated integrals.

Fubini's theorem begins with the single hypothesis that $\int_I f$ exists and deduces from it the existence of iterated integrals on I and their equality to $\int_I f$. From these primary consequences of integrability of f on I additional secondary deductions can be drawn. These are often valuable as necessary conditions for integrability. That is, their absence allows us to conclude nonexistence of $\int_I f$.

We have noted that there are always at least two iterated integrals for an interval I in $\overline{\mathbf{R}}^p$ with $p \geqslant 2$. It is necessary that all the iterated integrals of a given function f over I be equal if f is to be integrable over I. This is the fact which was used in Example 6 of Section 1.6 to show that $\int_I f$ does not exist when $I = [0, 1] \times [0, 1]$ and $f(x, y) = (x^2 - y^2)/(x^2 + y^2)^2$. (See p. 35.)

It is also necessary that each iterated integral possess the properties which the integral itself exhibits as a function of intervals. Two of the more obvious ones are existence of the iterated integral on every subinterval of I and additivity of the iterated integral. Note that mere existence of an iterated integral on I need not imply its existence on all subintervals. For instance, $\int_0^\infty \int_{-1}^1 x \, dx \, dy = 0$ but $\int_0^\infty \int_0^1 x \, dx \, dy$ does not exist. Thus existence of each iterated integral on all subintervals is another necessary condition for integrability which can be used to establish nonintegrability.

Additivity fails to be useful in this way. The following exercise explains why.

Exercise 1. Let $f: I \to \mathbf{R}^q$ possess an r-fold iterated integral on intervals G and H whose union is an interval. Show that this iterated integral also exists on $G \cup H$ and equals the sum of its values on G and H.

In summary, we have found two necessary conditions for integrability which are useful. They are *equality of iterated integrals* and *existence of iterated integrals on all subintervals*. Failure of either condition implies that the given function is not integrable.

Now we turn to sufficient conditions for integrability. One of the simplest ones to apply asserts absolute integrability. It makes use of measurable functions and the results of the preceding chapter.

TONELLI'S THEOREM.   *Let $f : I \rightarrow \mathbf{R}^q$ be measurable. Let $I = J \times K$. Suppose $|f| \leqslant g$ and $\int_K \int_J g(x, y) \, dx \, dy$ exists. Then f is absolutely integrable on I.*

It suffices to show that $\int_I |f|$ exists since the measurable function f will then be integrable by the comparison test. The function $|f|$ is also measurable. As we noted on page 139, there is a sequence of nonnegative integrable simple functions f_n which increases to $|f|$. Fubini's theorem applies to each f_n. Then $\int_J f_n(x, y) \, dx \leqslant \int_J g(x, y) \, dx$ for almost all y. Consequently

$$\int_I f_n = \int_K \int_J f_n(x, y) \, dx \, dy \leqslant \int_K \int_J g(x, y) \, dx \, dy$$

for all n. The boundedness of the sequence of integrals of f_n on I implies, by the monotone convergence theorem, that $\int_I |f|$ exists.

Two comments on Tonelli's theorem are in order. The function $|f|$ itself can be used in the role of g. Thus if $|f|$ possesses *one* iterated integral both f and $|f|$ are integrable. Contrast this with Example 6 of Section 1.6, where f has iterated integrals in both orders yet fails to be integrable. Of course f is not positive valued in that example. Secondly, note that f is the only function whose measurability needs to be checked. The measurability of $|f|$ follows from that of f. The only assumption on g is that an iterated integral exists.

Example 1. Let $I = [0, 1] \times [0, 1]$. Set $f(x, y) = |x - y|^{-\alpha}$ when $x \neq y$. Determine the values of α for which f is integrable on I.

Solution. Note first that the absence of a definition of f on the line segment $x = y$, $0 \leq x \leq 1$, is harmless since this segment is M-null. Since f is continuous elsewhere it is surely measurable.

Since this function is nonnegative it is enough to find the values of α for which one of the iterated integrals exists. The necessary condition and Tonelli's theorem together tell us that existence of an iterated integral and the double integral coincide. When $\alpha \geq 1$, there is no y in $[0, 1]$ for which $\int_0^1 |x - y|^{-\alpha} \, dx$ exists. Thus f is not integrable when $\alpha \geq 1$. When $\alpha < 1$, the appropriate primitives do exist and the iterated integral exists. The details are quite simple and need not be exhibited.

Example 2. Let $f_i : \mathbf{R} \to \mathbf{R}$ be continuous for $i = 1, 2, \ldots, p$. Suppose $|f_i| \leq g_i$ and $\int_{-\infty}^{\infty} g_i$ exists. Set $f(x_1, x_2, \ldots, x_p) = f_1(x_1)f_2(x_2) \cdots f_p(x_p)$. Show that f is integrable over $\overline{\mathbf{R}}^p$.

Solution. It is easy to show that f is continuous on $\mathbf{R}^p$, hence measurable. Define g from $g_1, g_2, \ldots, g_p$ as f is defined from the functions f_i. Then any p-fold integral of g exists. Hence f is integrable since $|f| \leq g$.

This example can be generalized by weakening the continuity assumption on the f_i. Another version is given in Exercise 10 at the end of the chapter.

There are also sufficient conditions which guarantee integrability without being confined to absolute integrability.

Above we explored the possibility of using the behavior of an iterated integral as a function of intervals to prove nonintegrability. That same behavior can also be used in a positive way to prove integrability. Instead of looking for misbehavior of an iterated integral which precludes integrability, we shall seek behavior which insures that Riemann sums approximate the iterated integral so closely that integrability follows. Here we have a parallel between the iterated integral in higher dimensions and the primitive in one dimension.

The type of argument for integrability which we propose to explore follows a pattern which has already become familiar. This pattern first appeared in the examples in Chapter 1. Here is how it looks in the present context.

Suppose $f : I \to \mathbf{R}^q$ has an iterated integral which is meaningful on every closed subinterval of I. Let $\phi(J)$ be the value of this iterated integral on a typical interval J. As noted in Exercise 1, the function ϕ is additive. Thus $\phi(I) = \phi(\mathfrak{D})$ for any division $\mathfrak{D}$ of I. This equation is crucial to the proof that $fM(\mathfrak{D})$ approximates $\phi(I)$ closely when $\mathfrak{D}$ is properly restricted.

Since a difference of sums is also a sum of differences, $f(z)M(J) - \phi(J)$ is the central object in an examination of $fM(\mathfrak{D}) - \phi(\mathfrak{D})$. Two types of behavior occur. Usually the difference $f(z)M(J) - \phi(J)$ is far closer to zero than the individual terms $f(z)M(J)$ and $\phi(J)$. The exception is to find these terms about as near zero as their difference.

Suppose that U is a subset of I such that the difference $f(z)M(J) - \phi(J)$ appears to be very small when $z \in U$. Let $X = I - U$. Each tagged division $\mathfrak{D}$ falls into disjoint subsets $\mathfrak{D}_U$ and $\mathfrak{D}_X$ whose tags are in U and X, respectively. The validity of the choice of U is established by defining γ on I so that $|(fM - \phi)(\mathfrak{D}_U)|$, $|fM(\mathfrak{D}_X)|$, and

$|\phi(\mathcal{D}_X)|$ are all small when $\mathcal{D}$ is γ-fine. (Keep in mind that $fM(\mathcal{D}) - \phi(I)$ can be expressed as $(fM - \phi)(\mathcal{D}_U) + fM(\mathcal{D}_X) - \phi(\mathcal{D}_X)$.)

The sums associated with $\mathcal{D}_U$ and $\mathcal{D}_X$ can all be estimated by passing to the larger sums of absolute values. Discussion of such calculations is facilitated by the introduction of appropriate language.

Definition. Let Φ be a function defined on tagged intervals. A set E is Φ-*null* provided there exists a gauge γ on E such that $|\Phi|(\mathcal{S}) < \epsilon$ whenever $\mathcal{S}$ is a γ-fine partial division whose tags all lie in E.

The outcome of the discussion above is as follows. *Suppose $f: I \to \mathbf{R}^q$ has an iterated integral $\phi(J)$ on each subinterval J of I. Suppose there is a set U such that U is $(fM - \phi)$-null and $I - U$ is both ϕ-null and fM-null. Then f is integrable on I and $\int_I f = \phi(I)$.*

It may seem that replacement of sums by the sums of absolute values is a risky proposition. Henstock's lemma assures us of the contrary in the circumstances we are considering. Indeed, by combining Fubini's theorem with Henstock's lemma we get an easy proof that $\int_I f$ exists only if I is $(fM - \phi)$-null.

Before going on with the discussion of iterated integrals let us note that the notion of Φ-null sets generalizes the earlier notion of M-null and fM-null sets. As in the special cases we can say that a subset of a Φ-null set is Φ-null and a countable union of Φ-null sets is also Φ-null. The proofs given for fM-null sets need only notational changes to adapt them to the general concept.

Even though we can in principle always take $U = I$ it is usually convenient to do otherwise. One of the ways to show that U is $(fM - \phi)$-null is to show that $|f(z)M(J) - \phi(J)| < \epsilon M(J)$. Continuity of f at z permits a fairly simple argument.

Exercise 2. Suppose f has an iterated integral ϕ on the closed subintervals of I. Let f be continuous at each point of U, $U \subseteq I$. Suppose U contains no infinite points. Show that U is $(fM - \phi)$-null.

From Exercise 2 we see that $\int_I f = \phi(I)$ *when f is continuous on a set U of finite points, ϕ exists on all closed subintervals of I, $I - U$ is fM-null, and $I - U$ is ϕ-null.*

An unbounded closed interval in a space of two or more dimensions has an uncountable set of infinite points. Thus $I - U$ is not countable unless I is bounded. When $I - U$ is countable it is automatically fM-null; to show that $I - U$ is ϕ-null it suffices to show that singleton sets $\{z\}$ are ϕ-null since a countable union of ϕ-null sets is also ϕ-null. It may happen that U can be chosen so that $I - U$ is finite, or even a single point.

Example 3. Let $I = [0, 1] \times [0, 1]$. Set $f(x, y) = (x + y)^{-\alpha}$ on I. Determine the values of α for which f is integrable.

Solution. When $J = [a, b] \times [c, d]$ set $\phi(J) = \int_c^d \int_a^b f(x, y)\, dx\, dy$. Since f is continuous on I except at $(0, 0)$ the crucial question is whether $\phi(J)$ exists when $a = c = 0$ and, if so, how it behaves. Large values of α are more likely than small ones to cause f to be nonintegrable. Thus we shall look first for large values of α for which $\phi(J)$ is not meaningful.

When $\alpha \neq 1$,

$$\int_0^b (x + y)^{-\alpha}\, dx = \frac{1}{1 - \alpha} (x + y)^{1 - \alpha} \Big|_{x = 0}^{b}$$

$$= \frac{1}{1 - \alpha} \left[(b + y)^{1 - \alpha} - y^{1 - \alpha} \right].$$

Clearly this function is not integrable on $[0, d]$ when $\alpha \geqslant 2$. Since the existence of $\phi(J)$ on all $J \subseteq I$ is necessary for existence of $\int_I f$, we conclude that $\int_I f$ does not exist when $\alpha \geqslant 2$.

When $1 < \alpha < 2$ and $J = [0, b] \times [0, d]$,

$$\phi(J) = (1 - \alpha)^{-1}(2 - \alpha)^{-1}\left[(b + d)^{2 - \alpha} - d^{2 - \alpha} - b^{2 - \alpha}\right].$$

Clearly $\phi(J)$ is small when b and d are sufficiently small. Thus $\{(0, 0)\}$ is ϕ-null and $\int_I f$ exists.

The proof of integrability in Example 3 could also have been accomplished by appeal to Tonelli's theorem. There would have been no significant reduction in the length of the proof, however. The next example illustrates the proof of integrability through the iterated integral when appeal to absolute integrability is not possible.

Example 4.   Let $\sum_{n=1}^{\infty} c_n$ be a convergent series. Define f on $[0, \infty) \times [0, \infty)$ as follows. Let $f(x, y) = c_n$ on $[n - 1, n) \times [n - 1, n)$ for $n = 1, 2, 3, \ldots$. Elsewhere set $f(x, y) = 0$. Show that f is integrable on $[0, \infty] \times [0, \infty]$ and that its integral equals the sum of the series.

Solution.   Figure 2 will help to keep the details straight.

It must be shown that one of the iterated integrals exists on every closed subinterval of $[0, \infty] \times [0, \infty]$. The symmetry in the definition of f means that $\int_c^d \int_a^b f(x, y) \, dx \, dy$ is as good a choice as the other. When at least one of $[a, b]$ and $[c, d]$ is bounded, the function f is an integrable step function on $[a, b] \times [c, d]$. Easy calculations of the double integral and the iterated integral show that the two of them are equal (without Fubini's theorem). Now let n be an integer and consider the interval $[n, \infty] \times [n, \infty]$. Clearly, $\int_n^{\infty} f(x, y) \, dx = c_k$

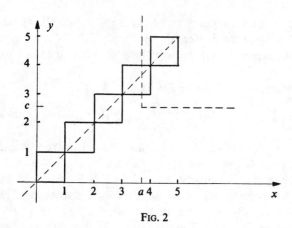

Fig. 2

when $k - 1 \leqslant y < k$ and $k - 1 \geqslant n$. Thus

$$\int_n^\infty \int_n^\infty f(x, y)\, dx\, dy = \sum_{k=n+1}^\infty c_k.$$

Let $[a, \infty] \times [c, \infty]$ be given. Select the smallest integer n such that $[n, \infty] \times [n, \infty] \subseteq [a, \infty] \times [c, \infty]$. Set $G = ([a, \infty] \times [c, \infty]) \cap ([n - 1, n] \times [n - 1, n])$. Separate consideration of the cases $n - 1 < a \leqslant n$ and $n - 1 < c \leqslant n$ shows that in both cases $\int_c^\infty \int_a^\infty f(x, y)\, dx\, dy = c_n M(G) + \int_n^\infty \int_n^\infty f(x, y)\, dx\, dy$.

Now let $\phi(J) = \int_c^d \int_a^b f(x, y)\, dx\, dy$ when $J = [a, b] \times [c, d]$. We will show next that $[0, \infty) \times [0, \infty)$ is $(fM - \phi)$-null and afterward that $\{\infty\} \times [0, \infty]$ and $[0, \infty] \times \{\infty\}$ are ϕ-null.

Let $I_n = [0, n] \times [0, n]$. Since $[0, \infty) \times [0, \infty) = \bigcup_{n=1}^\infty I_n$, it suffices to prove that each I_n is $(fM - \phi)$-null. We have noted that $\int_J f = \phi(J)$ when J is bounded. Recall also that integrability on an interval K implies that K is $(fM - \phi)$-null provided that $fM - \phi$ is restricted to subintervals of K. However, we want to know that I_n is $(fM - \phi)$-null when all partial divisions of $[0, \infty] \times [0, \infty]$ with tags in I_n are allowed, even if the intervals extend beyond I_n. All we need do to skirt this technical difficulty is to insist that $\gamma(z) \subseteq (-1, n + 1) \times (-1, n + 1)$ when $z \in I_n$. Then any γ-fine partial division of $[0, \infty] \times [0, \infty]$ with tags in I_n consists of subintervals of the larger closed interval I_{n+1}. Thus integrability of f on I_{n+1} implies that I_n is $(fM - \phi)$-null.

When z is an infinite point of $[0, \infty] \times [0, \infty]$ other than (∞, ∞), there is an interval $\gamma(z)$ on which f is everywhere zero. Thus $\phi(J) = 0$ when $z \in J$ and $J \subseteq \gamma(z)$. All that remains is to define $\gamma(\infty, \infty)$ so that $|\phi(J)| < \epsilon$ when $J = [a, \infty] \times [c, \infty]$ and $J \subseteq \gamma(\infty, \infty)$. Since $\phi(J) = c_n M(G) + \sum_{k=n+1}^{\infty} c_k$ with $M(G) \leqslant 1$, this is easily accomplished from the convergence of the series.

Of course the value of the integral of f on $[0, \infty] \times [0, \infty]$ is $\phi([0, \infty] \times [0, \infty])$, that is $\sum_{k=1}^{\infty} c_k$.

Tonelli's theorem has behind it many of the major concepts and results met so far. The resulting generality makes it simple to use in many cases. Sometimes it seems more appropriate to use means which rest on less sophisticated concepts and fewer major theorems. This is particularly true when dealing with continuous functions. The next exercise is offered in this spirit.

Exercise 3. Let $[a, b] \subseteq \mathbf{R}$. Let g and h be continuous on $[a, b]$ with $g \leqslant h$. Let E be the region bounded by the graphs of g and h. Let $f : E \to \mathbf{R}$ be continuous. Show that $\int_E f$ exists by using an iterated integral. Stick to the simplest arguments you can find.

Examples 1 and 2 can also be solved using less sophisticated tools than Tonelli's theorem. The price to be paid is a larger amount of computation directed toward showing that suitably chosen sets are ϕ-null.

S6.3. Compound divisions. Compatibility theorem.
The primary goal in this section is to prepare the ground
for the proof of Fubini's theorem. These preparations are
mainly concerned with gauges and divisions. The
principal result of this section is not only useful in proving
Fubini's theorem but also makes possible an easy
induction proof of the compatibility theorem for intervals
in $\bar{\mathbf{R}}^p$ with $p \geqslant 2$. Recall that it was shown in Section S1.8
that when $I = [a, b]$ and γ is a gauge on I there is a
division $\mathcal{D}$ of I which is γ-fine. In Section 1.6 we noted
that this proposition holds generally when $I \subseteq \bar{\mathbf{R}}^p$ for any
p but deferred the proof to this section.

Again let $\bar{\mathbf{R}}^p$ be factored into subspaces $\bar{\mathbf{R}}^t$ and $\bar{\mathbf{R}}^u$ by
grouping t one-dimensional subspaces to form $\bar{\mathbf{R}}^t$ and the
remaining u of them to form $\bar{\mathbf{R}}^u$. Express I in $\bar{\mathbf{R}}^p$ as
$G \times H$ with $G \subseteq \bar{\mathbf{R}}^t$ and $H \subseteq \bar{\mathbf{R}}^u$.

The main assertion in Fubini's theorem is that $\int_I f$
equals $\int_H \int_G f(x, y)\, dx\, dy$. Since the generalized Riemann
integral is defined in terms of Riemann sums we need to
relate Riemann sums which approximate $\int_I f$ to sums
which approximate the iterated integral. This is a
straightforward task which is easily accomplished by
looking at $\int_H \int_G f(x\ y)\, dx\, dy$ first.

The inner integration $\int_J f(x, y)\, dx$ is defined in terms of
the measure M_t on the subintervals of $\bar{\mathbf{R}}^t$. Thus
$\int_J f(x, y)\, dx$ is approximated by Riemann sums of the
form $f(\cdot, y) M_t(\mathcal{D}_y)$ where $\mathcal{D}_y$ is a division of G and $f(\cdot, y)$
is, as noted above, the function $x \to f(x, y)$.

For convenience set $h(y) = \int_G f(x, y)\, dx = \int_G f(\cdot, y)$.
The outer integration is the integral of h over H.
Consequently it is defined in terms of the interval measure
M_u on the subintervals of $\bar{\mathbf{R}}^u$. Thus $\int_H h$ can be
approximated by the Riemann sums of the form $h M_u(\mathcal{F})$
where $\mathcal{F}$ is a division of H. In other words, $h M_u(\mathcal{F})$
approximates $\int_H \int_G f(x, y)\, dx\, dy$.

A typical term of $hM_u(\mathcal{F})$ is $[\int_G f(\cdot, y)]M_u(K)$ for some $yK \in \mathcal{F}$. This is not appropriate to our purposes since it does not utilize values of f directly. Let's take the obvious step. It is to replace $\int_G f(\cdot, y)$ by its Riemann sum approximation $f(\cdot, y)M_t(\mathcal{D}_y)$. When this is done in every term of $hM_u(\mathcal{F})$ the result is this new approximation of the iterated integral:

$$\sum_{yK \in \mathcal{F}} \left[\sum_{xJ \in \mathcal{D}_y} f(x, y)M_t(J) \right] M_u(K).$$

Is this sum also a Riemann sum for $\int_I f$? Yes, it is. It is easy to see why. Distribute $M_u(K)$ over the inner sum. A typical term in the inner sum is then $f(x, y)M_t(J)M_u(K)$. But $M_t(J)M_u(K) = M(J \times K)$ where M is the interval measure in $\mathbf{R}^p$. Moreover the intervals $J \times K$ tagged by (x, y) form a "strip" partial division across I when yK is a fixed member of $\mathcal{F}$ and xJ ranges over $\mathcal{D}_y$. Let $\mathcal{D}$ be the union of all these partial divisions as yK ranges over $\mathcal{F}$ (i.e., $\mathcal{D}$ is the collection of all $(x, y)(J \times K)$ such that $xJ \in \mathcal{D}_y$ and $yK \in \mathcal{F}$). It is easy to verify that $\mathcal{D}$ is a tagged division of I. Moreover, the approximation to the iterated integral is just $fM(\mathcal{D})$.

Divisions of this special form will be called *compound* divisions. Figure 3 shows a compound division when $t = u = 1$. More precisely it is a compound of a set of divisions of G with a division of H. Figure 3 is an accurate depiction when $t = u = 1$. It can also be used as a schematic representation of a compound division for other values of t and u. To see how far this schematic diagram falls short of being faithful to its subject consider the cases $t = 2$, $u = 1$ and $t = 1$, $u = 2$. When $t = 2$, each $\mathcal{D}_y$ is a division of a rectangle G. The strip in Figure 3 corresponds to a slab cut into rectangular boxes. All of the tags for a single slab lie in a common plane. The stack

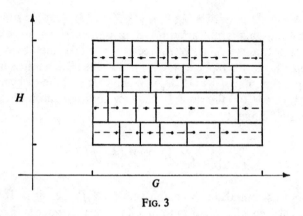

Fig. 3

of horizontal slabs constitutes $\mathcal{D}$. On the other hand, when $t = 1$, each $\mathcal{D}_y$ is a division of a line segment G. The counterpart of the strip in Figure 3 is a rectangular cylinder sliced into rectangular boxes by planes parallel to its ends. All the tags for the intervals making up a single cylinder lie on a line running through the cylinder from end to end. The division $\mathcal{D}$ is a bundle of such cylinders.

Let's return to the main task now. It is to learn how to approximate $\int_I f$ closely by $fM(\mathcal{D})$ when $\mathcal{D}$ is a compound division. This entails constructing the compound division $\mathcal{D}$ so that it is γ-fine for a given gauge γ on I. Instead of anticipating the result we will move forward step by step and then summarize what we have learned.

Each open interval in $\overline{\mathbf{R}}^p$ is a Cartesian product of open intervals in $\overline{\mathbf{R}}^t$ and $\overline{\mathbf{R}}^u$. Thus

$$\gamma(x, y) = \gamma_t(x, y) \times \gamma_u(x, y)$$

where $\gamma_t(x, y)$ is an open interval in $\overline{\mathbf{R}}^t$ and $\gamma_u(x, y)$ is an open interval in $\overline{\mathbf{R}}^u$. Moreover, $(x, y) \in \gamma(x, y)$ implies that $x \in \gamma_t(x, y)$ and $y \in \gamma_u(x, y)$. For each y the function $\gamma_t(\cdot, y): x \to \gamma_t(x, y)$ is a gauge on G. (The asymmetrical roles of G and H in the formation of $\mathcal{D}$ make $\gamma_u(x, \cdot)$ unimportant.) When $\mathcal{D}_y$ is $\gamma_t(\cdot, y)$-fine, $J \subseteq \gamma_t(x, y)$ for all $xJ \in \mathcal{D}_y$. We want to conclude that $J \times K \subseteq \gamma(x, y)$ when $xJ \in \mathcal{D}_y$ and y is the tag of K. Thus it is important that $K \subseteq \gamma_u(x, y)$ for every x which is a tag in $\mathcal{D}_y$. There are a finite number of elements in $\mathcal{D}_y$. Consequently, the intersection of all the open intervals $\gamma_u(x, y)$ as xJ varies over $\mathcal{D}_y$ is an open interval in $\overline{\mathbf{R}}^u$ which contains y. Set

$$\gamma'(y) = \bigcap_{xJ \in \mathcal{D}_y} \gamma_u(x, y). \qquad (*)$$

Now $J \times K \subseteq \gamma(x, y)$ when $xJ \in \mathcal{D}_y$ and $K \subseteq \gamma'(y)$.

These considerations enable us to make the following assertion.

Let γ be a gauge on $G \times H$. For each y in H let $\mathcal{D}_y$ be a division of G which is $\gamma_t(\cdot, y)$-fine. Let the gauge γ' be defined by $(*)$ for all y in H. Let $\mathcal{F}$ be a γ'-fine division of H. Then the compound division $\mathcal{D}$ formed from $\mathcal{F}$ and the divisions $\mathcal{D}_y$ for which y is a tag in $\mathcal{F}$ is γ-fine.

This proposition will be very useful in the proof of Fubini's theorem in the next section. It is also a natural tool in the proof of the existence of γ-fine divisions of intervals in $\overline{\mathbf{R}}^p$ with $p \geqslant 2$.

Example 5. Let $I \subseteq \overline{\mathbf{R}}^2$ and let γ be a gauge on I. Show that there is a γ-fine division $\mathcal{D}$ of I.

Solution. Let $I = G \times H$ with $G \subseteq \overline{\mathbf{R}}$ and $H \subseteq \overline{\mathbf{R}}$. From the compatibility theorem of Section S1.8 we know that there is a division $\mathcal{D}_y$ of G which is $\gamma_1(\cdot, y)$-fine for each y. Fix $\mathcal{D}_y$ for each y and form γ'. The same theorem assures the existence of a division $\mathcal{F}$ of H which is γ'-fine.

Now form the compound division $\mathcal{D}$ to get a γ-fine division of I.

The general case of an interval in $\overline{\mathbf{R}}^p$ can be handled similarly by using $t = 1$ and $u = p - 1$. Then the inductive hypothesis insures the existence of $\mathcal{F}$. Thus we have the following.

COMPATIBILITY THEOREM.   *Let I be a closed, nondegenerate interval in $\overline{\mathbf{R}}^p$. Let γ be a gauge on I. There is a γ-fine division of I.*

The discussion of compound divisions can be closed fittingly with a remark about doubly compound divisions. Figure 3 in Section 1.6 (p. 32) shows in (c) a division which is not compound. The more special case in (b) is a compound of three divisions of the horizontal interval with a division of the vertical interval. Part (a) is the most special of (a), (b), and (c). In (a) all three divisions of the horizontal interval are the same. Consequently it is also a compound of six divisions of the vertical interval with a division of the horizontal interval. Thus it is appropriate to say that (a) shows a doubly compound division. A natural question comes to mind at once. When γ is given on a two-dimensional interval, is there always a doubly compound division which is γ-fine? Exercise 7 at the end of this chapter shows that the answer is negative even if we do not require that the tags lie along lines in the intervals of the doubly compound division.

S6.4.   Proof of Fubini's theorem.   We assume $\int_I f$ exists where $I \subseteq \overline{\mathbf{R}}^p$ with $p \geqslant 2$. We adopt again the notations of Section S6.3, in particular $I = G \times H$. There are two things to be proved. First, it must be shown that $\int_G f(x, y)\, dx$ exists except for an M_u-null set of y's in H.

The Cauchy criterion is a natural tool to use in showing integrability since no candidate for the value of $\int_G f(x, y)\, dx$ is at hand. Second, we must show that $\int_H \int_G f(x, y)\, dx\, dy$ exists and equals $\int_I f$. The simplest procedure here is to show directly from the definition that $\int_I f$ is the integral of $\int_G f(x, \cdot)\, dx$ over H.

It is enough to prove the theorem for real-valued functions. For the remainder of the proof f is assumed to have real values.

Let N be the set of all y in H such that $\int_G f(x, y)\, dx$ does not exist. To show that N is M_u-null we may demonstrate the existence of a function $w : H \to \mathbf{R}$ such that $w(y) > 0$ when $y \in N$, $w(y) = 0$ when $y \in H - N$, and $\int_H w(y)\, dy = 0$. (Recall Exercise 2 of Section 4.1, p. 107.)

We begin by setting $w(y) = 0$ when $y \in H - N$. When $y \in N$ the value we need for $w(y)$ comes from the fact that the Cauchy criterion for the existence of $\int_G f(x, y)\, dx$ is violated. The negation of Cauchy's criterion reads as follows. For each y in N there exists $w(y) > 0$ such that, for every gauge γ_1 on G,

$$w(y) \leqslant |f(\cdot, y)M_t(\mathcal{D}_y) - f(\cdot, y)M_t(\mathcal{E}_y)|$$

for some γ_1-fine divisions $\mathcal{D}_y$ and $\mathcal{E}_y$ of G.

The appropriate choice of $\mathcal{D}_y$ and $\mathcal{E}_y$ requires a special γ_1. But whatever the divisions $\mathcal{D}_y$ and $\mathcal{E}_y$ may be, nothing more than an interchange of labels is needed to remove the absolute values. Thus we may suppose that

$$w(y) \leqslant f(\cdot, y)M_t(\mathcal{D}_y) - f(\cdot, y)M_t(\mathcal{E}_y) \qquad (**)$$

when $y \in N$. This inequality can be extended to all of H by choosing some γ_1-fine division $\mathcal{D}_y$ and setting $\mathcal{E}_y = \mathcal{D}_y$ when $y \in H - N$. (Then both sides of $(**)$ are zero when $y \in H - N$.)

Since the object is to show that $\int_H w = 0$, we seek a gauge γ' on H such that $wM_u(\mathfrak{F}) < \epsilon$ when $\mathfrak{F}$ is γ'-fine. To find γ' we must exploit our one hypothesis, the existence of $\int_I f$, and the technique for forming compound divisions which was developed in the preceding section.

For a given $\epsilon > 0$ there is a gauge γ on I such that $|fM(\mathfrak{D}) - fM(\mathfrak{E})| < \epsilon$ when $\mathfrak{D}$ and $\mathfrak{E}$ are γ-fine. As in the preceding section let $\gamma(x, y) = \gamma_t(x, y) \times \gamma_u(x, y)$. Use the gauge $\gamma_t(\cdot, y)$ in place of γ_1 and fix divisions $\mathfrak{D}_y$ and $\mathfrak{E}_y$ for which (**) holds. Do this for all y in H.

Through the procedure of the preceding section the family of divisions $\mathfrak{D}_y$ determines a gauge γ'_1 on H from γ. Similarly γ'_2 can be got using the divisions $\mathfrak{E}_y$ and the gauge γ. Let γ' be stricter than γ'_1 and γ'_2. Then any γ'-fine division $\mathfrak{F}$ of H yields γ-fine compound divisions $\mathfrak{D}$ and $\mathfrak{E}$ from the families $\mathfrak{D}_y$ and $\mathfrak{E}_y$, respectively.

Let $\mathfrak{F}$ be γ'-fine. Multiply both sides of (**) by $M_u(K)$ and sum over all yK in $\mathfrak{F}$. The result is $0 \leq wM_u(\mathfrak{F}) \leq fM(\mathfrak{D}) - fM(\mathfrak{E}) < \epsilon$ since $\mathfrak{D}$ and $\mathfrak{E}$ are γ-fine. Consequently $\int_H w = 0$.

Now set $h(y) = \int_G f(x, y)\, dx$ when $y \in H - N$ and extend h to all of H in some fashion, say by letting $h(y) = 0$ when $y \in N$. The remaining part of the proof is to show that $\int_H h = \int_I f$ by use of the definition of $\int_H h$.

This natural step looks promising:

$$\left| hM_u(\mathfrak{F}) - \int_I f \right| \leq \left| hM_u(\mathfrak{F}) - fM(\mathfrak{D}) \right| + \left| fM(\mathfrak{D}) - \int_I f \right|.$$

When $\mathfrak{D}$ is the compound of $\mathfrak{F}$ and divisions $\mathfrak{D}_y$, a rearrangement of sums yields

$$hM_u(\mathfrak{F}) - fM(\mathfrak{D})$$

$$= \sum_{yK \in \mathfrak{F}} \left[h(y) - f(\cdot, y)M_t(\mathfrak{D}_y) \right] M_u(K).$$

For convenience set $b(y) = h(y) - f(\cdot, y)M_t(\mathcal{D}_y)$. Then $|hM_u(\mathcal{F}) - fM(\mathcal{D})| \leq |b|M_u(\mathcal{F})$. The possibility of making $|b|M_u(\mathcal{F})$ small lies in two facts. First, $|b(y)|$ can be made small when $y \in H - N$ by a proper choice of $\mathcal{D}_y$. Second, the remaining terms arise from a partial division of H whose tags are in the M_u-null set N. (Our choices of $\mathcal{D}_y$ must also be made so that $fM(\mathcal{D})$ is close to $\int_I f$.)

Fix a gauge γ on I so that $|fM(\mathcal{D}) - \int_I f| < \epsilon$ when $\mathcal{D}$ is γ-fine. For each y in N fix some $\gamma_t(\cdot, y)$-fine division $\mathcal{D}_y$. For each y in $H - N$ choose $\mathcal{D}_y$ so that it is $\gamma_t(\cdot, y)$-fine and also so that $|b(y)| < \epsilon g(y)$ for a function g whose properties are specified in the next paragraph.

Let g be a strictly positive-valued function on H such that $gM_u(\mathcal{F}) \leq 1$ whenever $\mathcal{F}$ is γ_g-fine for a suitable gauge γ_g. Such a function has been constructed in the early paragraphs of Section S3.10. (See p. 96.)

The technique of the preceding section yields a gauge γ_1' on H from the divisions $\mathcal{D}_y$ and the gauge γ. Using the fact that N is M_u-null we fix γ_2 on N so that $|b|M_u(\mathcal{F}_N) < \epsilon$ when $\mathcal{F}_N$ is a γ_2-fine partial division with tags in N. Now let γ' be stricter than γ_1', γ_2, and γ_g.

Let $\mathcal{F}$ be γ'-fine. Let $\mathcal{F}_N$ be its subset whose tags are in N and let $\mathcal{G} = \mathcal{F} - \mathcal{F}_N$. Then $|b|M_u(\mathcal{F}_N) < \epsilon$ and $|b|M_u(\mathcal{G}) < \epsilon g M_u(\mathcal{G})$. Since $\mathcal{G} \subseteq \mathcal{F}$ and terms in $gM_u(\mathcal{F})$ are nonnegative we can also say that $gM_u(\mathcal{G}) \leq gM_u(\mathcal{F}) \leq 1$. Now

$$|b|M_u(\mathcal{F}) = |b|M_u(\mathcal{G}) + |b|M_u(\mathcal{F}_N) < 2\epsilon.$$

The choices of γ_1' and the divisions $\mathcal{D}_y$ also insure that $\mathcal{D}$ is γ-fine; consequently $|fM(\mathcal{D}) - \int_I f| < \epsilon$. In summary, we have $|hM_u(\mathcal{F}) - \int_I f| < 3\epsilon$ and therefore $\int_H h = \int_I f$.

S6.5. Double series. This section is an aside. It is not essential for the remaining chapters.

A double series is to a function on $N \times N$ as a series is to a function on N. That is, it is a family of partial sums and a limit process applied to them. The simplest way to proceed is to use rectangular sums. Let a_{ij} be the value of the function at (i, j). Let $S_{mn} = \sum_{j=1}^{n} \sum_{i=1}^{m} a_{ij}$. The limit of S_{mn} as m and n get large is one of the possible sums of the double series. A less obvious way to assign a sum is to use the generalized Riemann integral on $N \times N$. The assignment of the sum through the generalized Riemann integral has the distinct advantage that many of the theorems we have already proved are available for use.

The definition of the generalized Riemann integral on $\overline{N} \times \overline{N}$ parallels the definition on $\overline{R}^2$. That is, for divisions and gauges we use Cartesian products of intervals in $\overline{N}$. Thus $\gamma(\infty, \infty) = [r, \infty] \times [s, \infty]$, $\gamma(i, \infty) = \{i\} \times [s_i, \infty]$, $\gamma(\infty, j) = [r_j, \infty] \times \{j\}$, and $\gamma(i, j) = \{(i, j)\}$ when i and j are finite. A division $\mathcal{D}$ of $\overline{N}^2$ which is γ-fine has an unbounded interval $[t + 1, \infty] \times [u + 1, \infty]$ tagged with (∞, ∞), u intervals $[1 + c_j, \infty] \times \{j\}$, and t intervals $\{i\} \times [1 + d_i, \infty]$. The bounded intervals all consist of one point and their union contains a largest interval $[1, m] \times [1, n]$.

If we let $f(i, j)$ be another symbol for a_{ij}, the Riemann sum $fM(\mathcal{D})$ will have among its terms all those terms which appear in S_{mn}. But $fM(\mathcal{D})$ will also include other terms. For instance, when $m = t$ and $n = u$ it contains the terms for the points in $[1 + m, c_j] \times \{j\}$, $1 \leq j \leq n$, and $\{i\} \times [1 + n, d_i]$, $1 \leq i \leq m$. Thus the existence of the integral is not equivalent to existence of the limit of rectangular sums. (Keep in mind that $M(\{i, j\}) = 1$ when i and j are finite and M vanishes on all unbounded intervals.) The reward for using the more complicated limit process is, of course, the opportunity to use the properties of the generalized Riemann integral.

In particular, it is possible to apply Fubini's theorem. When the double series converges in the generalized Riemann sense, Fubini's theorem asserts that $\sum_{i=1}^{\infty} a_{ij}$ and $\sum_{j=1}^{\infty} a_{ij}$ exist for all i and j in $\mathbf{N}$ and

$$\sum_{i,j} a_{ij} = \sum_{j=1}^{\infty} \sum_{i=1}^{\infty} a_{ij} = \sum_{i=1}^{\infty} \sum_{j=1}^{\infty} a_{ij}.$$

The monotone convergence theorem and the results on integration on subsets are also available. One notable simplification is present in $\overline{\mathbf{N}}^2$. Integrability on bounded sets is automatic for any function on $\mathbf{N}^2$. Consequently, the integrability of a nonnegative function on $\overline{\mathbf{N}}^2$ is equivalent to boundedness of its rectangle sums. Absolute integrability can be determined by comparison in many instances.

Exercises 12–16 pursue double series a bit further.

6.6. Exercises.

4. Let $f(x, y) = 1$ on every interval of the form $[i, i + 1) \times [i - 1, i)$ and $f(x, y) = -1$ on every interval of the form $[i - 1, i) \times [i, i + 1)$ where $i = 1, 2, 3, \ldots$. Let $f(x, y) = 0$ otherwise. Show that f has iterated integrals on subintervals of $[0, \infty] \times [0, \infty]$ but that f is not integrable.

5. Let $f(x, y) = x/y$ when $y > 0$ and $f(x, 0) = 0$. Show that $\int_0^1 \int_{-1}^1 f(x, y)\, dx\, dy$ exists but that $\int_0^1 \int_0^1 f(x, y)\, dx\, dy$ does not exist.

6. (a) Let $f(x, y) = (-1)^{i+j}(ij)^{-1}$ when $(x, y) \in [i - 1, i] \times [j - 1, j)$ for all positive integers i and j. Is f integrable on $[0, \infty] \times [0, \infty]$?
  (b) Replace ij by $i + j$ in (a). Same question.

7. Let $I = [0, 1] \times [0, 1]$. Let A be the subset of I consisting of all points $(1/i, 0)$ and $(0, 1/i)$ for $i = 1, 2, 3, \ldots$. Let γ be a gauge on I such that $\gamma(z) \cap A = \{z\}$ when $z \in A$ and $\gamma(z) \cap A$ is empty when $z \notin A \cup \{(0, 0)\}$. Show that there is no γ-fine division of I which consists of rows and columns of intervals, in the manner of Figure 3(a) of Section 1.6.

8. Let $f(x, y) = 1$ when x is rational and $f(x, y) = y$ when x is irrational.

(a) Determine where f is continuous.

(b) Let ϕ be one of the iterated integrals of f. Show that $\mathbf{R}^2$ is $(\phi - fM)$-null.

(c) Show that f is integrable on each closed and bounded interval.

9. Give an alternative proof of nonintegrability of the function in Example 6, Section 1.6, by showing that $\{(0, 0)\}$ fails to be ϕ-null where ϕ is one of the iterated integrals on the subintervals of $[0, 1] \times [0, 1]$.

10. Let $f_i : \mathbf{R} \to \mathbf{R}$ be measurable for $i = 1, 2, \ldots, p$. Suppose $|f_i| \leqslant g_i$ and $\int_{-\infty}^{\infty} g_i$ exists. Set $f(x_1, x_2, \ldots, x_p) = f_1(x_1) \cdot f_2(x_2) \cdots f_p(x_p)$. Show that f is absolutely integrable on $\overline{\mathbf{R}}^p$.

11. Give an alternative proof of Fubini's theorem for absolutely integrable functions using approximation by step functions and/or approximation by simple functions.

12. Suppose $\sum_{i, j} a_{ij}$ exists in the generalized Riemann sense.

(a) Show that the rectangle sums S_{mn} have $\sum_{i, j} a_{ij}$ as limit as m and n tend to infinity.

(b) Form triangle sums $T_n = \sum_{i + j \leqslant n} a_{ij}$. Show that $\lim_{n \to \infty} T_n$ exists and equals the double sum also.

13. Let $\sum_{i=1}^{\infty} b_i$, $\sum_{j=1}^{\infty} c_j$, and $\sum_{j=1}^{\infty} |c_j|$ all be convergent. Set $a_{ij} = b_i c_j$. Show that $\sum_{i, j} a_{ij}$ exists in the generalized Riemann sense.

14. Under the assumptions of Exercise 13 show that

$$\left(\sum_{i=1}^{\infty} b_i \right) \left(\sum_{j=1}^{\infty} c_j \right) = \sum_{n=2}^{\infty} \sum_{i + j = n} b_i c_j.$$

15. Let $a_{ij} = 0$ when $i = j$ and $a_{ij} = (-1)^{i+j}/(i-j)$ when $i \neq j$. Does $\sum_{i,j} a_{ij}$ exist in the generalized Riemann sense?

16. Is there a double sequence a_{ij} such that

$$\sum_{i=1}^{\infty} \sum_{j=1}^{\infty} a_{ij} = \sum_{j=1}^{\infty} \sum_{i=1}^{\infty} a_{ij}$$

but $\sum_{i,j} a_{ij}$ does not exist in the generalized Riemann sense?

CHAPTER 7

INTEGRALS OF STIELTJES TYPE

The Stieltjes integral is a generalization of the Riemann integral on intervals of real numbers. It employs a function α in place of the identity function in forming the measure of subintervals of $[a, b]$. A Riemann sum $f\Delta\alpha(\mathcal{D})$ is made up of terms $f(z)(\alpha(v) - \alpha(u))$. The resulting integral is denoted $\int_a^b f \, d\alpha$ or $\int_a^b f(x) \, d\alpha(x)$. The function α is called the integrator and f is the integrand.

The level of generality of $\int_a^b f \, d\alpha$ depends on the limit process which is applied to the Riemann sum. There are three which should claim our attention. The least general of them is the limit used customarily in the Riemann integral. The second one makes use of the notion of refinement of divisions. The third is the one used in the generalized Riemann integral. These three limits, and their associated integrals, will be designated by the names *norm*, *refinement*, and *gauge*. The integrals will be distinguished as $(\mathcal{N})\int_a^b f \, d\alpha$, $(\mathcal{R})\int_a^b f \, d\alpha$, and $(\mathcal{G})\int_a^b f \, d\alpha$ wherever this is essential.

The principal focus of the chapter will be on the gauge integral. Sometimes it is helpful to draw the stronger conclusion that the refinement or norm integral exists. Unless a contrary indication is made the gauge integral is intended.

When comparing this treatment with other sources, keep in mind that the norm and refinement limits are the ones which are used elsewhere.

The first section contains detailed statements of the three definitions. It also contains the first of the examples which show how the interplay of properties of integrand and integrator determines existence of the integral.

The second and third sections give a concise account of the elementary properties of the Stieltjes integral. These include linearity in integrand and integrator, additivity over intervals, inequalities, component-wise integration, conversion of a Stieltjes integral to a Riemann integral, and limit properties of functions defined as integrals. Most of the proofs of these properties are familiar from calculus courses or from earlier chapters.

Section 7.4 is given over to examples. They are largely a preparation for the existence theorems in the following section. They show that certain combinations of properties of integrand and integrator are needed for existence of integrals. The existence theorems deal with well-known classes of functions. Various combinations of step functions, monotone functions, functions of bounded variation, and regulated functions are used as integrands and integrators.

Integration by parts is a valuable tool for the evaluation of integrals. In Stieltjes integrals it also plays a role in the completion of the existence theorems. This comes about because $\int_a^b f \, d\alpha$ and $\int_a^b \alpha \, df$ are related in integration by parts. Thus restrictions on integrand and integrator can be interchanged. Several levels of generality in the properties of integrand and integrator are considered in Section 7.6. Under severe restrictions the integration by parts equation expresses $\int_a^b f \, d\alpha$ in terms of $\int_a^b \alpha \, df$ and the values of f and α at the endpoints. For more general f and α there must be other terms to account for the effect of common right-hand and left-hand discontinuities of f and α. This version of integration by parts has not arisen in discussions limited to norm and refinement integrals, since such common discontinuities prevent the existence of norm and refinement integrals.

The criterion for absolute integrability in Section 7.7 is a generalization of the one in Chapter 3. Here, as there, the integrability of $|f|$ is determined by whether the indefinite integral $\int_a^x f \, d\alpha$ is a function of bounded variation. The most obvious assumption upon which this question can be attacked is that the integrators be increasing functions. It is possible, however, to allow α to be a function of bounded variation. If so, $|f|$ must be integrated with respect to the variation function of α rather than α itself. When α is increasing this is the same as integrating $|f|$ against α.

The discussion of lattice operations is restricted to increasing integrators. It is simply a transcription of the results of Section 3.3 with appropriate notational changes.

The monotone convergence theorem is a notational variant of the one in Section 3.5. In it the integrator is an increasing function. The dominated convergence theorem, on the other hand, is general enough to allow functions of bounded variation as integrators. This is possible because the proof given in Section S3.10 deals with dominated convergence on its own terms. The same argument is valid for Stieltjes integrals with respect to functions of bounded variation, thanks to the absolute integrability results of Section 7.7.

The change of variables results in Section 7.9 are given for several sets of hypotheses on f, α, and the function which is being substituted. Some of the results are closely related to those given in Section 2.7.

In Section 7.10 a mean value theorem for Stieltjes integrals is made the source of mean value theorems for Riemann integrals of products of functions. Some sufficient conditions for existence of integrals of products are obtained in passing.

Sections S7.11–S7.14 are designated as optional on first reading. The first two of them give some additional facts about Stieltjes integrals. The last two sections complete arguments which were deferred from earlier sections of this chapter.

7.1. Three versions of the Riemann-Stieltjes integral.
In the integral $\int_a^b f \, d\alpha$ it is not essential that f and α both have real values, though that is the most important case. It is essential that there be a product of values of f and $\Delta\alpha$. The product $f(z)\Delta\alpha(J)$ has the necessary properties when f has values in $\mathbf{R}$ and α has values in $\mathbf{R}^q$ or the other way around. (The notation $f(z)\Delta\alpha(J)$ will be maintained when f has vector values even though this violates the usual convention of placing the scalar factor on the left.) It is also possible to let f and α have complex values and form $f(z)\Delta\alpha(J)$ in the sense of complex number multiplication. Unless a restriction is mentioned all these possibilities will be allowed in subsequent statements.

The device which made possible the definition of integrals on unbounded intervals also permits the definition of $\int_a^b f \, d\alpha$ when α is defined on one of the intervals $(a, b]$, $[a, b)$, or (a, b) instead of on $[a, b]$. When α is defined on $[u, v]$ let $\Delta\alpha([u, v]) = \alpha(v) - \alpha(u)$. When $[u, v] \subseteq [a, b]$ but at least one endpoint of $[u, v]$ is not in the domain of α, set $\Delta\alpha([u, v]) = 0$. It follows that the value of f at an endpoint of $[a, b]$ where α is not defined does not influence $\int_a^b f \, d\alpha$.

The three limit processes which produce three levels of generality of $\int_a^b f \, d\alpha$ are the following.

Let $[a, b]$ be a bounded interval. Let $\mathcal{D}$ be a division of $[a, b]$. The *norm* of $\mathcal{D}$, denoted $\|\mathcal{D}\|$, is the maximum of the lengths $L(J)$ as J ranges over $\mathcal{D}$. Then A is an $\mathcal{N}$-limit of $f\Delta\alpha(\mathcal{D})$ provided that for each positive ϵ there exists a positive δ such that $|A - f\Delta\alpha(\mathcal{D})| < \epsilon$ whenever $\|\mathcal{D}\| < \delta$.

The refinement limit, briefly $\mathcal{R}$-limit, is formulated as follows. For each positive ϵ there exists a division $\mathcal{F}$ of $[a, b]$ such that $|A - f\Delta\alpha(\mathcal{D})| < \epsilon$ whenever $\mathcal{D}$ is a refinement of $\mathcal{F}$. (Recall that $\mathcal{D}$ is a refinement of $\mathcal{F}$ when each interval of $\mathcal{F}$ is a union of intervals of $\mathcal{D}$.)

The gauge limit, or $\mathcal{G}$-limit, of $f\Delta\alpha(\mathcal{D})$ is A when $|A - f\Delta\alpha(\mathcal{D})| < \epsilon$ for all γ-fine $\mathcal{D}$.

Exercise 1. Show that if A is the $\mathcal{N}$-limit of $f\Delta\alpha$ on $[a, b]$ then A is also the $\mathcal{R}$-limit of $f\Delta\alpha$ on $[a, b]$. Likewise, if A is the $\mathcal{R}$-limit of $f\Delta\alpha$ on $[a, b]$, then A is the $\mathcal{G}$-limit of $f\Delta\alpha$ on $[a, b]$.

When f is constant, say $f(x) = K$ for all x in $[a, b]$, $f\Delta\alpha(\mathcal{D})$ telescopes to $K(\alpha(b) - \alpha(a))$ provided $[a, b]$ is the domain of α. Consequently $\int_a^b f \, d\alpha = K(\alpha(b) - \alpha(a))$ in all three senses whatever α may be. On the other hand, when α is constant on $[a, b]$ every term in $f\Delta\alpha(\mathcal{D})$ is zero and $\int_a^b f \, d\alpha = 0$ for any f.

When neither f nor α is constant on $[a, b]$, the interplay of the properties of f and α dictates which versions of $\int_a^b f \, d\alpha$ exist.

Example 1. Let α be defined on $[a, b]$. Fix c in the open interval (a, b). Let $f(x) = 1$ when $a \leqslant x \leqslant c$ and $f(x) = 0$ when $c < x \leqslant b$. Examine each of the versions of $\int_a^b f \, d\alpha$.

Solution. The terms $f(z)\Delta\alpha(J)$ are zero for those tags z which lie in $(c, b]$. Thus $f\Delta\alpha(\mathfrak{D})$ telescopes to $\alpha(t) - \alpha(a)$ where t is the right endpoint of the last interval of $\mathfrak{D}$ whose tag is in $[a, c]$. Each type of limit restricts t in its relation to c in a particular way. These restrictions on t determine the properties of α which correspond to existence of the integral in its three versions. We shall examine them in turn.

Consider first the $\mathcal{G}$-limit. There is no loss of generality in requiring that $c \notin \gamma(z)$ when $z \neq c$. When $\mathfrak{D}$ is γ-fine each interval of $\mathfrak{D}$ which contains c has c as its tag. Thus $t \in \gamma(c)$. Since t is also the left endpoint of the first interval of $\mathfrak{D}$ whose tag is in $(c, b]$ we can also claim that $c < t$. Moreover, for each t such that $c < t$ and $t \in \gamma(c)$, there is a γ-fine $\mathfrak{D}$ for which t is the right endpoint of the last interval of $\mathfrak{D}$ whose tag is in $[a, c]$. Since $\gamma(c)$ can be an arbitrarily small interval containing c, it is now clear that existence of $(\mathcal{G})\int_a^b f\, d\alpha$ is equivalent to existence of $\alpha(c +)$, i.e., the right-hand limit of α at c. Moreover, $\int_a^b f\, d\alpha = \alpha(c +) - \alpha(a)$.

Next consider the $\mathcal{R}$-limit. This time we may insist that each division $\mathcal{F}$ have c among the endpoints of its intervals. When $\mathfrak{D}$ is a refinement of $\mathcal{F}$ the number c is among the endpoints of the intervals of $\mathfrak{D}$ also. Thus $c \leqslant t$ and moreover t is in the interval of $\mathcal{F}$ having c as its left endpoint. Conversely, given any t in the interval of $\mathcal{F}$ having c as left endpoint, there is a tagged division $\mathfrak{D}$ which is a refinement of $\mathcal{F}$ and satisfies $f\Delta\alpha(\mathfrak{D}) = \alpha(t) - \alpha(a)$. Thus the existence of the $\mathcal{R}$-limit of $f\Delta\alpha(\mathfrak{D})$ is equivalent to right-hand continuity of α at c and $\int_a^b f\, d\alpha = \alpha(c) - \alpha(a)$.

The final case is the $\mathcal{N}$-limit. Let δ be a positive number satisfying $a < c - \delta$ and $c + \delta < b$. Suppose $\|\mathfrak{D}\| < \delta$. It follows that $c - \delta < t < c + \delta$. Conversely, given t in $(c - \delta, c + \delta)$, there is a tagged division $\mathfrak{D}$ for which $\|\mathfrak{D}\| < \delta$ and $f\Delta\alpha(\mathfrak{D}) = \alpha(t) - \alpha(a)$. Thus existence of the $\mathfrak{N}$-limit of $f\Delta\alpha(\mathfrak{D})$ is equivalent to continuity of α at c and, of course, $\int_a^b f\, d\alpha = \alpha(c) - \alpha(a)$.

Note that the discontinuity of f on the right at c in this example calls forth continuity of α at c in order for the norm integral to exist. However, right-hand continuity of α at c is enough for existence of the refinement integral. Existence of the gauge integral requires no continuity at c. A right-hand limit suffices in this case.

In Exercise 1 it was noted that the refinement integral is at least as general as the norm integral and the gauge integral is at least as general as the refinement integral. Example 1 shows that these relations are actually strict; i.e., no two of the integrals are equivalent.

7.2. Basic properties of Riemann-Stieltjes integrals. For a given f, α, and interval $[a, b]$ each of the limit processes determines at most one value for the integral. Moreover, each of the limits has a Cauchy criterion which may be used to determine existence of the limit even when the value of the integral cannot be identified. (The proofs of the uniqueness and Cauchy criterion assertions are the same as the ones given in earlier chapters.)

The Cauchy criterion can be used as in Section 2.3 to show that existence of $\int_a^b f\, d\alpha$ in one of the three senses implies existence of the integral in the same sense on each closed subinterval of $[a, b]$.

The linearity of the integral as a function of the integrand, i.e.,

$$\int_a^b (f + g)\, d\alpha = \int_a^b f\, d\alpha + \int_a^b g\, d\alpha$$

and

$$\int_a^b (cf)\, d\alpha = c \int_a^b f\, d\alpha,$$

is proved just as it was in the case where α is the identity function.

There is a new linearity property for the Riemann-Stieltjes integral as a function of the integrator. If $\int_a^b f\, d\alpha$ and $\int_a^b f\, d\beta$ exist and c is constant, then f is integrable with respect to $\alpha + \beta$ and $c\alpha$ and

$$\int_a^b f\, d(\alpha + \beta) = \int_a^b f\, d\alpha + \int_a^b f\, d\beta,$$

$$\int_a^b f\, d(c\alpha) = c \int_a^b f\, d\alpha.$$

In both forms of linearity the constant c may be complex when the functions have complex values. Moreover, induction extends the linearity to finite linear combinations.

The proof of linearity in the integrator differs little from the proof of linearity in the integrand. There is no need to set the proof down.

Additivity over intervals,

$$\int_a^b f\, d\alpha = \int_a^c f\, d\alpha + \int_c^b f\, d\alpha$$

when $a < c < b$, requires the same kind of proof as in Section 2.4. Again it is true that integrability over $[a, c]$ and $[c, b]$ implies integrability over $[a, b]$ with the integral being additive. Moreover, an induction argument extends additivity to finite additivity over any division of $[a, b]$.

When one of f and α is vector-valued and the other is real-valued, integration can be carried out component by component. That is, the jth component of $\int_a^b f\,d\alpha$ is $\int_a^b f\,d\alpha_j$ when $\alpha = (\alpha_1, \alpha_2, \ldots, \alpha_q)$ and it is $\int_a^b f_j\,d\alpha$ when $f = (f_1, f_2, \ldots, f_q)$. Moreover existence of $\int_a^b f\,d\alpha$ is equivalent to existence of all of the component integrals. All these assertions follow readily from the fact that limits of vector-valued functions may be calculated component by component.

When f and α are both complex-valued, the facts are a little more complicated. Let $f = f_1 + if_2$ and $\alpha = \alpha_1 + i\alpha_2$ with all of f_1, f_2, α_1, and α_2 real-valued. Then

$$f\Delta\alpha = (f_1\Delta\alpha_1 - f_2\Delta\alpha_2) + i(f_2\Delta\alpha_1 + f_1\Delta\alpha_2).$$

Existence of $\int_a^b f\,d\alpha$ is equivalent to existence of the limits of the components, i.e., $f_1\Delta\alpha_1 - f_2\Delta\alpha_2$ and $f_2\Delta\alpha_1 + f_1\Delta\alpha_2$. In place of this awkward necessary and sufficient condition one may use a neater sufficient condition. The linearity of the integral in both the integrator and the integrand allows us to say that $\int_a^b f\,d\alpha$ exists when all of $\int_a^b f_1\,d\alpha_1$, $\int_a^b f_1\,d\alpha_2$, $\int_a^b f_2\,d\alpha_1$, and $\int_a^b f_2\,d\alpha_2$ exist. Moreover,

$$\int_a^b f\,d\alpha = \int_a^b f_1\,d\alpha_1 - \int_a^b f_2\,d\alpha_2 + i\left(\int_a^b f_2\,d\alpha_1 + \int_a^b f_1\,d\alpha_2\right).$$

The order properties of the Riemann-Stieltjes integral require hypotheses on both the integrand and the integrator. Suppose $0 \leqslant g$ and α is increasing on its domain, whether $[a, b]$ or another interval with endpoints a and b. Then $g\Delta\alpha$ is nonnegative on every tagged division of $[a, b]$. Consequently $0 \leqslant \int_a^b g\,d\alpha$. More generally, when $f \leqslant g$ and α is increasing, $\int_a^b f\,d\alpha \leqslant \int_a^b g\,d\alpha$.

Next suppose $|f| \leqslant g$ and $|\Delta\alpha| \leqslant \Delta\beta$. Then $\left|\int_a^b f\,d\alpha\right| \leqslant \int_a^b g\,d\beta$ when both integrals are meaningful. Note that f and α need not be real-valued. Moreover, no claim is made yet about integrals of $|f|$. The proof needed for this inequality is essentially the same as the one called for in Exercise 2 of Section 2.1. (See p. 250.)

All of the properties of the integral given so far in this section apply equally well to all three versions of the Riemann-Stieltjes integral. The next one is claimed only for the gauge integral, however,

Suppose $\alpha(x) = \alpha(a) + \int_a^x g\, d\beta$ for all x in $[a, b]$. Then $(\mathcal{G})\int_a^b f\, d\alpha$ exists if and only if $(\mathcal{G})\int_a^b fg\, d\beta$ exists. Moreover, $\int_a^b f\, d\alpha = \int_a^b fg\, d\beta$.

Note that when β is the identity function this proposition reduces integrals of Stieltjes type to ordinary integrals. Thus it makes the fundamental theorem available for the evaluation of Stieltjes integrals. Clearly, such a reduction is possible when α is a primitive, i.e., when α is continuous on $[a, b]$ and $\alpha' = g$ except possibly on a countable subset of $[a, b]$.

Exercise 2. Give a proof of the above proposition expressing $\int_a^b f\, d\alpha$ as $\int_a^b fg\, d\beta$.

The solution of Exercise 2 requires Henstock's lemma in the version appropriate to Riemann-Stieltjes integrals. Henstock's lemma will also be needed in the next section. The Riemann-Stieltjes version comes readily from the one stated on page 74. The proof is the same as the one given in Section S3.7.

HENSTOCK'S LEMMA. *Suppose $\int_a^b f\, d\alpha$ exists. Let γ be a gauge on $[a, b]$ such that $|f\Delta\alpha(\mathcal{D}) - \int_a^b f\, d\alpha| < \epsilon$ when $\mathcal{D}$ is any γ-fine division of $[a, b]$. Let $\mathcal{E}$ be a subset of a γ-fine division of $[a, b]$. Then $|\sum[f\Delta\alpha(z[u, v]) - \int_u^v f\, d\alpha]| \leqslant \epsilon$ and $\sum|f\Delta\alpha(z[u, v]) - \int_u^v f\, d\alpha| \leqslant 2q\epsilon$ when the sums are taken* *over all $z[u, v]$ in $\mathscr{E}$ and q is the dimension of the range space of $f\alpha$.*

7.3. Limits, continuity, and differentiability of integrals.

The first question to be taken up is whether $\int_a^b f \, d\alpha$ exists when $\lim_{t \to b} \int_a^t f \, d\alpha$ exists. The type of argument which resolves such a question is like the ones used earlier in discussing improper integrals. We need not repeat the argument here.

Suppose $\int_a^t f \, d\alpha$ exists for all t in (a, b). Suppose also that $\lim_{t \to b} f(b) \, \Delta \alpha \, ([t, b])$ exists. Then $\int_a^b f \, d\alpha$ exists if and only if $\lim_{t \to b} \int_a^t f \, d\alpha$ exists. Moreover,

$$\int_a^b f \, d\alpha = \lim_{t \to b} \int_a^t f \, d\alpha + \lim_{t \to b} f(b) \Delta \alpha ([t, b]).$$

Note that when α is defined only on $[a, b)$, and consequently we set $\Delta \alpha([t, b]) = 0$, the conclusion is that $\int_a^b f \, d\alpha = \lim_{t \to b} \int_a^t f \, d\alpha$. Of course this is a generalization of what we found earlier in considering $\int_a^\infty f(x) \, dx$. On the other hand, when α is defined on $[a, b]$ and $\alpha(b-)$ exists, the conclusion is that $\int_a^b f \, d\alpha = \lim_{t \to b} \int_a^t f \, d\alpha + f(b)$ $\cdot (\alpha(b) - \alpha(b-))$.

The corresponding limit statement for the left endpoint is

$$\int_a^b f \, d\alpha = \lim_{s \to a} \int_s^b f \, d\alpha + \lim_{s \to a} f(a) \Delta \alpha ([a, s]).$$

The full statement of the left endpoint proposition is just like the right endpoint statement above.

A number of other "expanding interval" limit problems can be reduced to these two.

Now consider the limit properties of the function F defined on an interval containing c by the equation $F(x) = \int_c^x f\, d\alpha$. Here we suppose the integral is an oriented integral so that

$$F(y) - F(x) = \int_x^y f\, d\alpha$$

whatever the relation of y to x.

Suppose α and F are defined on an interval containing z. Using Henstock's lemma we can find $\gamma(z)$ such that $|\int_x^y f\, d\alpha - f(z)(\alpha(y) - \alpha(x))| < \epsilon$ when $z \in [x, y]$ and $[x, y] \subseteq \gamma(z)$. From this we can determine when F has one-sided limits. In fact, $F(z+)$ exists when $f(z)(\alpha(y) - \alpha(z))$ has a limit as y approaches z from the right and

$$F(z+) - F(z) = \lim_{y \to z^+} f(z)(\alpha(y) - \alpha(z)).$$

Similarly,

$$F(z-) - F(z) = \lim_{y \to z^-} f(z)(\alpha(y) - \alpha(z)).$$

The relation of a limit of $f(z)(\alpha(y) - \alpha(z))$ to the one-sided limit of α depends on whether $f(z)$ is zero. At any rate

$$F(z+) - F(z) = f(z)(\alpha(z+) - \alpha(z))$$

and similarly for left-hand limits when α has one-sided limits at z. Consequently, *F is continuous at z when α is continuous at z.*

Now we ask about the derivative of F. When α is the identity function, we know that $F'(z) = f(z)$ when f is continuous at z. This can be generalized but strong restrictions on α are needed to keep the proof elementary.

Suppose α *is increasing on an interval containing* z, f *is continuous at* z, F *exists on an interval containing* z, *and* $\alpha'(z)$ *exists. Then* $F'(z) = f(z)\alpha'(z)$.

In a sufficiently small interval on which α is increasing $|\int_x^y [f(t) - f(z)]\, d\alpha(t)| \leq \epsilon |\alpha(y) - \alpha(x)|$. With this estimate and the equation

$$F(y) - F(x) - f(z)(\alpha(y) - \alpha(x))$$
$$= \int_x^y [f(t) - f(z)]\, d\alpha(t)$$

we can easily pass to the conclusion $F'(z) = f(z)\alpha'(z)$.

7.4. Values of certain integrals. The specific integrals which are evaluated in this section illustrate some of the facts of the preceding two sections. They will be helpful in obtaining more general results in the next section, too.

Example 2. Evaluate $\int_a^b x^i\, dx^j$ with $i > -j$ and $j > 0$.

Solution. Since $\alpha(x) = x^j$ defines a function which is continuous on $[a, b]$ and differentiable except possibly at $x = 0$, it suffices to evaluate $\int_a^b jx^{i+j-1}\, dx$. Since $i + j > 0$ the fundamental theorem applies. Thus

$$\int_a^b x^i\, dx^j = \frac{j}{i+j}\, x^{i+j}\bigg|_a^b = \frac{j}{i+j}\, (b^{i+j} - a^{i+j}).$$

A start on the integration of step functions was made in Example 1 above (p. 181). We need to complete the discussion but only for the $\mathcal{G}$-integral. The next example is another step toward the general case.

Example 3. Let $c \in [a, b]$. Let $f(c) = 1$ and $f(x) = 0$ otherwise. Show that $(\mathcal{G})\int_a^b f \, d\alpha$ exists if and only if α has one-sided limits at c and that $\int_a^b f \, d\alpha = \alpha(c+) - \alpha(c-)$. (We set $\alpha(a-) = \alpha(a)$ and $\alpha(b+) = \alpha(b)$.)

Solution. It is not difficult to apply the definition here. However the results of the previous section are quite convenient.

First let $c = b$. Then $\int_a^t f \, d\alpha = 0$ when $a < t < b$. Consequently, since $f(b) = 1$, $\int_a^b f \, d\alpha$ exists if and only if $\alpha(b-)$ exists and $\int_a^b f \, d\alpha = \alpha(b) - \alpha(b-)$.

Now let $c = a$. A similar argument shows that $\int_a^b f \, d\alpha = \alpha(a+) - \alpha(a)$.

When $a < c < b$, we consider $\int_a^c f \, d\alpha$ and $\int_c^b f \, d\alpha$ separately, using the preceding cases. Then additivity gives the desired conclusion.

The remaining special step function we need to consider is one which is zero except on an open interval.

Example 4. Let $(c, d) \subseteq [a, b]$. Let $f(x) = 1$ when $x \in (c, d)$ and $f(x) = 0$ otherwise. Show that $(\mathcal{G})\int_a^b f \, d\alpha$ exists if and only if $\alpha(c+)$ and $\alpha(d-)$ exist and that $\int_a^b f \, d\alpha = \alpha(d-) - \alpha(c+)$.

Solution. Since f is zero on $[a, c]$ and on $[d, b]$ we have the immediate reduction $\int_a^b f \, d\alpha = \int_c^d f \, d\alpha$. Let $c < e < d$. Consider $\int_c^e f \, d\alpha$ and $\int_e^d f \, d\alpha$. Since $f(c) = 0$, the limit expression for $\int_c^e f \, d\alpha$ reduces to $\lim_{s \to c} \int_s^e f \, d\alpha$. But $\int_s^e f \, d\alpha = \alpha(e) - \alpha(s)$ since $f(x) = 1$ for all x in $[s, e]$. Thus $\int_c^e f \, d\alpha$ exists if and only if $\alpha(c+)$ exists. Moreover $\int_c^e f \, d\alpha = \alpha(e) - \alpha(c+)$. For similar reasons $\int_e^d f \, d\alpha = \alpha(d-) - \alpha(e)$. By addition $\int_a^b f \, d\alpha = \alpha(d-) - \alpha(c+)$.

Every step function is a linear combination of functions like those in Examples 3 and 4. Thus, *if f is a step function* on $[a, b]$ *and* α *has one-sided limits at every discontinuity of* f, *then* $(\mathcal{G})\int_a^b f\,d\alpha$ *exists.*

Example 3 by itself is enough to show us that $(\mathcal{G})\int_a^b f\,d\alpha$ exists for a fixed α and all step functions f only if α has one-sided limits everywhere in $[a, b]$.

The next exercise records the general expressions for the integrals $\int_a^b f\,d\alpha$ and $\int_a^b \alpha\,df$ when f is a step function.

Exercise 3. Let $a = x_0 < x_1 < \cdots < x_n = b$. Let f be a step function on $[a, b]$ with $f(x) = F_j$ when $x_{j-1} < x < x_j$.

(a) Suppose α has one-sided limits at $x_0, x_1, \ldots, x_n$. For convenience set $\alpha(a-) = \alpha(a)$ and $\alpha(b+) = \alpha(b)$. Show that

$$\int_a^b f\,d\alpha = \sum_{j=1}^n F_j(\alpha(x_j -) - \alpha(x_{j-1} +))$$

$$+ \sum_{j=0}^n f(x_j)(\alpha(x_j +) - \alpha(x_j -)).$$

(b) Show that $\int_a^b \alpha\,df$ exists for any function α and that

$$\int_a^b \alpha\,df = \sum_{j=0}^{n-1} \alpha(x_j)(F_{j+1} - f(x_j)) + \sum_{j=1}^n \alpha(x_j)(f(x_j) - F_j).$$

An increasing function has one-sided limits at each point in its domain. Thus a step function can be integrated with respect to a function which is increasing on $[a, b]$.

The next example shows that even when α is defined on $[a, b)$ and increasing with $\lim_{x \to b} \alpha(x) = \infty$ there are positive functions which can be integrated with respect to α.

Example 5. Let $\alpha : [a, b) \to \mathbf{R}$ be increasing. Let $a = a_0 < a_1 < a_2 < \cdots$ with $\lim_{n \to \infty} a_n = b$. Let $(F_n)_{n=1}^\infty$

be a decreasing sequence of constants satisfying $F_n(\alpha(a_n) - \alpha(a_{n-1})) \leqslant 1/2^n$ for $n = 1, 2, 3, \ldots$. Set $f(x) = F_n$ when $a_{n-1} \leqslant x < a_n$. Show that $\int_a^b f \, d\alpha$ exists.

Solution. The function f is a step function on each $[a, t]$ with $t < b$. Thus $\int_a^t f \, d\alpha$ exists. From Section 7.3 we know that existence of $\int_a^b f \, d\alpha$ is equivalent to existence of $\lim_{t \to b} \int_a^t f \, d\alpha$. Since f is positive and α is increasing, $\int_a^t f \, d\alpha$ increases as t increases. Thus it is enough to find an upper bound. Now $f(x) \leqslant F_n$ when $x \in [a_{n-1}, a_n]$. Thus $\int_s^t f \, d\alpha \leqslant F_n(\alpha(t) - \alpha(s)) \leqslant 1/2^n$ when $a_{n-1} \leqslant s \leqslant t \leqslant a_n$. For a given t let n be chosen so that $a_{n-1} \leqslant t < a_n$. From the additivity of the integral, $\int_a^t f \, d\alpha \leqslant \sum_{i=1}^n 1/2^i < 1$.

With the aid of Exercise 3(b) we can connect series convergence and integration in another way which is sometimes quite useful.

Exercise 4. For $j = 0, 1, 2, \ldots$ let $f(x) = j$ when $j \leqslant x < j + 1$. Let α be any function defined on $[0, \infty)$. Show that $\int_0^\infty \alpha \, df$ exists if and only if $\sum_{j=1}^\infty \alpha(j)$ is convergent and that

$$\int_0^\infty \alpha \, df = \sum_{j=1}^\infty \alpha(j).$$

7.5. Existence theorems for Riemann-Stieltjes integrals. Certain combinations of properties of f and α imply the existence of $\int_a^b f \, d\alpha$. A little information of this kind has been obtained in the preceding section. Now we shall go on to more general results.

Throughout this section it will be assumed that α is defined on a *closed* interval.

The classes of functions which are to be used as integrators and integrands are the step functions, continuous functions, monotone functions, functions of bounded variation, and the regulated functions. Of these function classes only the last has not been given a formal definition.

Definition. A function f is a *regulated* function on $[a, b]$ provided f has a right-hand limit and a left-hand limit at each point of $[a, b]$.

As usual we shall use $f(x +)$ for the right-hand limit at x and $f(x -)$ for the left-hand limit. The convention that $f(a -) = f(a)$ and $f(b +) = f(b)$ will be in force.

The regulated functions include all the other classes mentioned above. It is evident that step functions and continuous functions are regulated functions. To show that a monotone function has right-hand and left-hand limits is an exercise in the application of the definition of least upper bound and greatest lower bound. To show that a function of bounded variation is a regulated function requires a bit more effort. The details may be found in Section S7.13. (See p. 222.)

One of the conclusions from the preceding section has the following consequence. *When f is a step function and α is a regulated function $(\mathcal{G})\int_a^b f \, d\alpha$ exists.* To get more general results from this we shall use uniformly convergent sequences of step functions. First we need to know what functions can be approximated uniformly by step functions.

A function f is a regulated function on $[a, b]$ if and only if f is the uniform limit of a sequence of step functions.

A proof of this proposition is given in Section S7.13. (See p. 224.)

We also need to formulate a uniform convergence theorem for Riemann-Stieltjes integrals.

Suppose $(f_n)_{n=1}^{\infty}$ converges uniformly to f on $[a, b]$ and α has bounded variation on $[a, b]$. Suppose $\int_a^b f_n \, d\alpha$ exists for each n. Then $\lim_{n\to\infty}\int_a^b f_n \, d\alpha$ and $\int_a^b f \, d\alpha$ exist and are equal.

The proof required here is essentially the same as the one given in Section 3.4, page 84. It should be noted that the same type of integral is used in the conclusion as in the hypotheses. For instance, if $(\mathfrak{N})\int_a^b f_n \, d\alpha$ exists for each n, then $(\mathfrak{N})\int_a^b f \, d\alpha$ exists.

Our main result is now evident from the foregoing propositions. *Suppose f is a regulated function and α is a function of bounded variation on $[a, b]$. Then $(\mathcal{G})\int_a^b f \, d\alpha$ exists.*

By placing stronger restrictions on f we can claim more about the sense in which the integral exists. The principal existence theorem for the norm integral is the following. *Suppose f is a continuous function and α is a function of bounded variation on $[a, b]$. Then $(\mathfrak{N})\int_a^b f \, d\alpha$ exists.*

The customary proof of this proposition uses the uniform continuity of f and the Cauchy criterion for the existence of $(\mathfrak{N})\int_a^b f \, d\alpha$. There should be no difficulty in supplying the details.

In the next section we will see that the hypotheses on f and α in the two existence theorems above can be interchanged. That is, *suppose f is a function of bounded variation on $[a, b]$. Then $(\mathcal{G})\int_a^b f \, d\alpha$ exists when α is a regulated function and $(\mathfrak{N})\int_a^b f \, d\alpha$ exists when α is a continuous function.*

When f and α are both regulated functions but neither has bounded variation $\int_a^b f \, d\alpha$ may or may not exist.

Exercise 5. Construct regulated functions f and α such that $\int_a^b \alpha \, d\alpha$ exists but $\int_a^b f \, d\alpha$ does not.

7.6. Integration by parts. The integration by parts formula for Riemann integrals arises from the derivative formula for products. Since $(f\alpha)' = f\alpha' + f'\alpha$, it follows that $\int_a^b f\alpha' = \int_a^b (f\alpha)' - \int_a^b f'\alpha$ under suitable hypotheses. The fundamental theorem then gives the usual equation:

$$\int_a^b f\alpha' = f(b)\alpha(b) - f(a)\alpha(a) - \int_a^b f'\alpha.$$

We noted in Section 2.1 that it suffices that f and α be primitives. (Recall that a primitive on $[a, b]$ is continuous on $[a, b]$ and differentiable except possibly on a countable subset.) In Section 7.2 we found that $\int_a^b f\alpha'$ and $\int_a^b f'\alpha$ can be converted into Riemann-Stieltjes integrals. Thus we have the following.

Suppose f and α are primitives on $[a, b]$. Then $\int_a^b f\,d\alpha$ exists if and only if $\int_a^b \alpha\,df$ exists. Moreover, when these integrals exist,

$$\int_a^b f\,d\alpha = f(b)\alpha(b) - f(a)\alpha(a) - \int_a^b \alpha\,df.$$

Since this equation makes no explicit use of derivatives, it is quite natural to seek other conditions on f and α under which the integration by parts formula holds. The most direct way to proceed is to attempt to recast the Riemann sum for $\int_a^b f\,d\alpha$ as a sum for $\int_a^b d(f\alpha) - \int_a^b \alpha\,df$. (Keep in mind that $\int_a^b d(f\alpha) = f(b)\alpha(b) - f(a)\alpha(a)$ for any functions f and α.) We will assume $\int_a^b \alpha\,df$ exists and have as our goal the existence of $\int_a^b f\,d\alpha$ and the integration by parts formula. Consequently, it will be necessary to consider Riemann sums $f\Delta\alpha(\mathcal{D})$ in full generality. On the other hand, since $\int_a^b \alpha\,df$ is assumed to exist, we may use sums $\alpha\Delta f(\mathcal{D})$ of special form whenever it proves expedient to do so.

Let us consider a typical term in $f\Delta\alpha(\mathcal{D})$. Some simple algebra gives

$$f(z)(\alpha(v) - \alpha(u)) = f(v)\alpha(v) - f(u)\alpha(u)$$
$$- \left[\alpha(u)(f(z) - f(u)) + \alpha(v)(f(v) - f(z))\right].$$

The bracketed expression is part of a Riemann sum $\alpha\Delta f(\mathcal{D}')$ for a tagged division $\mathcal{D}'$ associated with $\mathcal{D}$. The tagged interval $z[u, v]$ of $\mathcal{D}$ gives rise to one or two intervals of $\mathcal{D}'$ as follows. When $z = u$ let $v[u, v]$ be in $\mathcal{D}'$. When $z = v$ let $u[u, v]$ be in $\mathcal{D}'$. When $u < z < v$, let both $u[u, z]$ and $v[z, v]$ be in $\mathcal{D}'$. Now it follows from the equation above that $f\Delta\alpha(\mathcal{D}) = \Delta(f\alpha)(\mathcal{D}) - \alpha\Delta f(\mathcal{D}')$.

Note that no interval of $\mathcal{D}'$ has the same tag as the interval of $\mathcal{D}$ which contains it. Thus $\mathcal{D}'$ need not be γ-fine when $\mathcal{D}$ is γ-fine. Consequently, $\mathcal{D}'$ is of no use for the $\mathcal{G}$-integral. On the other hand, $\mathcal{D}'$ is a refinement of $\mathcal{D}$. Consequently $\|\mathcal{D}'\| \leqslant \|\mathcal{D}\|$. This means that $\mathcal{D}'$ is useful in the $\mathcal{R}$-integral and the $\mathcal{N}$-integral. Indeed, it is clear that the following is true.

Suppose that $(\mathcal{R})\int_a^b \alpha \, df$ exists. Then $(\mathcal{R})\int_a^b f \, d\alpha$ also exists and $\int_a^b f \, d\alpha = f(b)\alpha(b) - f(a)\alpha(a) - \int_a^b \alpha \, df$. Moreover the same assertion holds with $\mathcal{R}$-integrals replaced by $\mathcal{N}$-integrals.

This proposition makes good the claim near the end of the previous section that $(\mathcal{N})\int_a^b f \, d\alpha$ exists when α is continuous and f is a function of bounded variation.

The result above does not carry over to $\mathcal{G}$-integrals. We can draw from our previous stock of examples to get a counterexample.

Example 6. Let $f(a) = 1$ and $f(x) = 0$ when $a < x \leqslant b$. Show that existence of $\int_a^b \alpha \, df$ does not imply existence of

$\int_a^b f\,d\alpha$ and that the integration by parts formula need not hold when both integrals exist.

Solution. In Exercise 3, part (b), $(\mathcal{G})\int_a^b \alpha\,df$ was found to exist for every α. Furthermore $\int_a^b \alpha\,df = -\alpha(a)$. On the other hand, $(\mathcal{G})\int_a^b f\,d\alpha$ exists if and only if $\alpha(a+)$ exists. Moreover $\int_a^b f\,d\alpha = \alpha(a+) - \alpha(a)$. Consequently, when both integrals exist,

$$\int_a^b f\,d\alpha = f(b)\alpha(b) - f(a)\alpha(a)$$

$$-\int_a^b \alpha\,df + \alpha(a+) - \alpha(a).$$

Example 6 has shown us that, for $\mathcal{G}$-integrals, the integration by parts formula needs a correction term. Since the equation $f\Delta\alpha(\mathcal{D}) = \Delta(f\alpha)(\mathcal{D}) - \alpha\Delta f(\mathcal{D}')$ can be rewritten as

$$f\Delta\alpha(\mathcal{D}) = \Delta(f\alpha)(\mathcal{D}) - \alpha\Delta f(\mathcal{D}) + \left[\alpha\Delta f(\mathcal{D}) - \alpha\Delta f(\mathcal{D}')\right],$$

it is from $\alpha\Delta f(\mathcal{D}) - \alpha\Delta f(\mathcal{D}')$ that we should seek the correction term. Let us examine the terms in this expression arising from $z[u,v]$ in $\mathcal{D}$. Addition and subtraction of $f(z)$ in $f(v) - f(u)$ followed by regrouping gives

$$\alpha(z)(f(v) - f(u)) - \left[\alpha(u)(f(z) - f(u))\right.$$
$$\left. + \alpha(v)(f(v) - f(z))\right]$$
$$= (\alpha(z) - \alpha(u))(f(z) - f(u))$$
$$- (\alpha(z) - \alpha(v))(f(z) - f(v)).$$

For convenience set $\Phi(z[u,v]) = (\alpha(z) - \alpha(u))(f(z) - f(u))$ and $\Psi(z[u,v]) = (\alpha(z) - \alpha(v))(f(z) - f(v))$. With this notation the earlier equation for $f\Delta\alpha(\mathcal{D})$ becomes

$$f\Delta\alpha(\mathcal{D}) = \Delta(f\alpha)(\mathcal{D}) - \alpha\Delta f(\mathcal{D}) + \Phi(\mathcal{D}) - \Psi(\mathcal{D}).$$

When the $\mathcal{G}$-limit of $\Phi(\mathcal{D}) - \Psi(\mathcal{D})$ exists there is a correction term for the integration by parts formula. This proves to be the case when one of f and α is a function of bounded variation and the other is a regulated function.

Consider Φ first. Always $u \leqslant z$. When $u = z$, $\Phi(z[u, v])$ $= 0$. When $u < z$ and $[u, v] \subseteq \gamma(z)$ the value of $\Phi(z[u, v])$ is close to $(\alpha(z) - \alpha(z-))(f(z) - f(z-))$. Thus the $\mathcal{G}$-limit of $\Phi(\mathcal{D})$ appears to be a sum of terms $\phi(z)$ where

$$\phi(z) = (\alpha(z) - \alpha(z-))(f(z) - f(z-)).$$

Similarly one may conjecture that the $\mathcal{G}$-limit of $\Psi(\mathcal{D})$ is a sum of $\psi(z)$ where

$$\psi(z) = (\alpha(z) - \alpha(z+))(f(z) - f(z+)).$$

The value of $\phi(z)$ is zero except where α and f are both discontinuous from the left. Similarly $\psi(z)$ is zero except where α and f are both discontinuous from the right. Each regulated function has only a countable set of discontinuities. (The reasons for this are given in Section S7.13, p. 225.) The assumption that one of f and α has bounded variation and the other is regulated is clearly more than enough to assure that there is an infinite series $\sum_{n=1}^{\infty}\phi(c_n)$ which accounts for all nonzero values of ϕ. A similar statement can be made about ψ.

The same assumptions on f and α guarantee absolute convergence of each of these series. Suppose f has bounded variation. The regulated function α is bounded since it is the uniform limit of bounded functions, namely step functions. Consequently $|\phi(z)| \leqslant K|f(z) - f(z-)|$ for some constant K. Thus absolute convergence of any series of values $\phi(c_n)$ follows from convergence of

$\sum_{n=1}^{\infty}|f(c_n) - f(c_n-)|$. The convergence of this series is established in Section S7.13, page 222. Of course similar assertions can be made about ψ.

The essential elements have been identified but some detailed arguments have yet to be made.

Exercise 6. Let f be a function of bounded variation and α a regulated function on $[a, b]$. Let $c_1, c_2, c_3, \ldots$ be distinct points in $[a, b]$ including all points where ϕ is not zero. Show that $\sum_{n=1}^{\infty}\phi(c_n)$ is the $\mathcal{G}$-limit of $\Phi(\mathcal{D})$.

Now we can give the integration by parts formula with the correction term.

Suppose one of f and α is a function of bounded variation and the other is a regulated function. Then $(\mathcal{G})\int_a^b f \, d\alpha$ and $(\mathcal{G})\int_a^b \alpha \, df$ exist. Moreover

$$\int_a^b f \, d\alpha = f(b)\alpha(b) - f(a)\alpha(a)$$

$$- \int_a^b \alpha \, df + \sum_{n=1}^{\infty} (\phi(c_n) - \psi(c_n))$$

where $c_1, c_2, c_3, \ldots$ are distinct points including all points where f and α are simultaneously discontinuous from the right and all points where they are simultaneously discontinuous from the left.

Here is an illustration of the application of this formula.

Example 7. Suppose $(a_n)_{n=1}^{\infty}$ is a strictly decreasing sequence in $(0, 1)$ with $\lim_{n\to\infty} a_n = 0$. Suppose $\sum_{n=1}^{\infty}|t_n| < \infty$. Set $f(0) = 0$. When $0 < x \leqslant 1$ set $f(x) = \sum_{n=k}^{\infty} t_n$ where k is the first integer such that $a_k \leqslant x$. Evaluate $\int_0^1 f \, df$.

Solution. The function f is constant on each interval $[a_{n+1}, a_n)$ and $f(a_n) - f(x) = t_n$ when $a_{n+1} \leqslant x < a_n$. From this fact and the convergence of $\sum_{n=1}^{\infty} |t_n|$ we conclude that f is a function of bounded variation. Moreover $f(z) = f(z-) = f(z+)$ when z is in $(0, 1]$ and z is not an a_n. Also $f(a_n) = f(a_n +)$ but $f(a_n) - f(a_n -) = t_n$. Since the tail of a convergent series goes to zero, $f(0) = f(0 +)$. Consequently, $\sum_{n=1}^{\infty} \phi(a_n) = \sum_{n=1}^{\infty} t_n^2$ and $\psi(z) = 0$ for all z. Integration by parts gives

$$\int_0^1 f \, df = \frac{1}{2} \left(\sum_{n=1}^{\infty} t_n \right)^2 + \frac{1}{2} \sum_{n=1}^{\infty} t_n^2.$$

This integral can be evaluated directly from the definition, too. The result is $\int_0^1 f \, df = \sum_{n=1}^{\infty} t_n \sum_{k=n}^{\infty} t_k$. Consequently, for any absolutely convergent series,

$$\sum_{n=1}^{\infty} \sum_{k=n}^{\infty} t_n t_k = \frac{1}{2} \left(\sum_{n=1}^{\infty} t_n \right)^2 + \frac{1}{2} \sum_{n=1}^{\infty} t_n^2.$$

7.7. Integration of absolute values. Lattice operations. When α is the identity function, the criterion for the integrability of $|f|$ goes as follows. Suppose $\int_a^b f$ exists. Set $F(x) = \int_a^x f$. Then $\int_a^b |f|$ exists if and only if F is a function of bounded variation. Moreover $\int_a^b |f| = \text{lub}_{\mathscr{D}} |\Delta F|(\mathscr{D})$.

For more general integrators α the obvious question is whether one should expect to get $\int_a^b |f| \, d\alpha = \text{lub}_{\mathscr{D}} |\Delta F|(\mathscr{D})$ when $F(x) = \int_a^x f \, d\alpha$. If α is an increasing function, the integral of $|f|$ with respect to α is at least the right kind of thing when it exists, i.e., a nonnegative real number. When α is not increasing or not real-valued, $\int_a^b |f| \, d\alpha$ is not the right sort of object. In this case a new integrator must be put in place of α. The appropriate one is the variation function associated with α.

Definition. Let $\alpha : [a, b] \rightarrow \mathbf{R}^q$ be a function of bounded variation. The *total variation* of α on $[a, b]$, denoted $V_a^b \alpha$, is given by

$$V_a^b \alpha = \operatorname*{lub}_{\mathcal{D}} |\Delta \alpha|(\mathcal{D}).$$

When α has bounded variation on $[a, b]$, it also has bounded variation on every subinterval of $[a, b]$. Thus it is possible to consider $V_a^x \alpha$ for every x such that $a < x \leqslant b$. By adopting the convention that $V_a^a \alpha = 0$, we may define the *variation function* of α on $[a, b]$ by the rule $x \rightarrow V_a^x \alpha$.

The total variation is additive. That is, when $a < c < b$, $V_a^b \alpha = V_a^c \alpha + V_c^b \alpha$. (See Section S7.13, p. 221, for a proof.) Since the total variation is clearly nonnegative on every interval, this means that the variation function is an increasing function. Thus we are sure that $\int_a^b |f(x)|\, dV_a^x \alpha$ is at least the right sort of integral for our purposes. The next example shows that the obvious generalization is sound when α is severely restricted.

Example 8. Let α be a step function on $[a, b]$. Show that $\int_a^b |f(x)|\, dV_a^x \alpha = V_a^b F$ where $F(x) = \int_a^x f\, d\alpha$.

Solution. Let $a = x_0 < x_1 < \cdots < x_n = b$ be such that $\alpha(x)$ has the constant value A_j when $x_{j-1} < x < x_j$. Then $\int_u^v f\, d\alpha = 0$ when $[u, v]$ is contained in some (x_{j-1}, x_j). Consequently, F is also a step function which is constant on each interval (x_{j-1}, x_j). Moreover, from Exercise 3,

$$F(x) - F(x_{j-1}) = f(x_{j-1})(A_j - \alpha(x_{j-1}))$$

and

$$F(x_j) - F(x) = f(x_j)(\alpha(x_j) - A_j)$$

when $x_{j-1} < x < x_j$.

It is easy to see from the definition of total variation that the total variation of a step function is the sum of the absolute values of the jumps of the function. Consequently,

$$V_a^b F = \sum_{j=1}^{n} |f(x_{j-1})||A_j - \alpha(x_{j-1})| + \sum_{j=1}^{n} |f(x_j)||\alpha(x_j) - A_j|.$$

A similar calculation gives us $V_a^x \alpha$. What we actually need, however, is the jumps in $V_a^x \alpha$ at the points x_j. For convenience let $V_a^x \alpha$ have the value V_j when $x_{j-1} < x < x_j$ and the value v_j when $x = x_j$. Now

$$V_j - v_{j-1} = |A_j - \alpha(x_{j-1})|$$

and

$$v_j - V_j = |\alpha(x_j) - A_j|$$

for $j = 1, 2, \ldots, n$.

From Exercise 3 we have the formula

$$\int_a^b |f(x)| \, dV_a^x \alpha = \sum_{j=0}^{n-1} |f(x_j)|(V_{j+1} - v_j)$$

$$+ \sum_{j=1}^{n} |f(x_j)|(v_j - V_j).$$

When we introduce the values for $V_{j+1} - v_j$ and $v_j - V_j$, we get a sum equivalent to the one for $V_a^b F$.

It is possible to go well beyond the special circumstances of Example 8.

ABSOLUTE INTEGRABILITY THEOREM. *Suppose $\int_a^b f \, d\alpha$ exists and α is a function of bounded variation on $[a, b]$. Set $F(x) = \int_a^x f \, d\alpha$. Then $\int_a^b |f(x)| \, dV_a^x \alpha$ exists if and only if F is a function of bounded variation. Moreover, $V_a^b F = \int_a^b |f(x)| dV_a^x \alpha$.*

The techniques used in Section S3.8 for the earlier absolute integrability proof are necessary to prove this theorem, too. Something more is needed, however. The approximation of sums containing terms $|f(z)| V_u^v \alpha$ by sums of $|f(z)||\alpha(v) - \alpha(u)|$ requires a version of Henstock's lemma for the $\mathcal{R}$-limit. The complete proof may be found in Section S7.14.

This version of the absolute integrability theorem does not apply to $\int_a^\infty f(x)\, dx$, since the identity function does not have bounded variation on $[a, \infty]$. The absolute integrability theorem in Section 3.2 did not exclude unbounded intervals. Thus one might expect a further extension to be valid. It can be obtained as a corollary of the above theorem provided one uses the relation between $V_a^b F$ and $\lim_{t \to b} V_a^t F$. This latter fact is given in Section S7.13, page 223.

Exercise 7. Extend the absolute integrability theorem to a function α which has $[a, b)$ as its domain and which is a function of bounded variation on $[a, t]$ for all $t < b$.

As a very easy corollary of the absolute integrability criterion we have the following comparison test.

Suppose $\int_a^b f\, d\alpha$ and $\int_a^b g(x)\, dV_a^x \alpha$ exist and $|f| \leqslant g$. Then $\int_a^b |f(x)|\, dV_a^x \alpha$ also exists.

When α is increasing on its domain $[a, b]$ or $[a, b)$, the variation function satisfies $V_a^x \alpha = \alpha(x) - \alpha(a)$. Thus integration with respect to the variation function is the same as integration with respect to α, since a constant term in the integrator makes no contribution to the integral. Thus the above comparison test can be specialized as follows.

Suppose α is increasing on its domain $[a, b)$ or $[a, b]$. Suppose $\int_a^b f\, d\alpha$ and $\int_a^b g\, d\alpha$ exist and $|f| \leqslant g$. Then $\int_a^b |f|\, d\alpha$ also exists.

This comparison test is the crucial fact in the extension of the lattice operations, as given in Section 3.3, to integrals with increasing integrators. The arguments given there carry over without change to yield the next proposition.

Let α be increasing on its domain $[a, b)$ or $[a, b]$. Let $\int_a^b h\, d\alpha$ and $\int_a^b f_i\, d\alpha$ exist for $i = 1, 2, \ldots, n$. Suppose $f_i \leqslant h$ for all i or $h \leqslant f_i$ for all i. Then $\bigwedge_{i=1}^n f_i$ and $\bigvee_{i=1}^n f_i$ are integrable with respect to α on $[a, b]$.

As a convenience in this section, only two of the intervals having endpoints a and b have been mentioned. Corresponding propositions for $(a, b]$ and (a, b) are valid. Those for $(a, b]$ require only notational changes from the corresponding ones for $[a, b)$. Once both of these are in hand, the open interval (a, b) can be dealt with by expressing it as the union of $(a, c]$ and $[c, b)$ for any c between a and b.

7.8. Monotone and dominated convergence.

The monotone convergence theorem can be formulated for an integrator which is defined on one of the intervals with endpoints a and b and is increasing on its domain.

MONOTONE CONVERGENCE THEOREM. *Let $\int_a^b f_n\, d\alpha$ exist for $n = 1, 2, 3, \ldots$. Let α be increasing on its domain and let $(f_n)_{n=1}^\infty$ be monotone on $[a, b]$ with limit f. Then $\int_a^b f\, d\alpha$ exists if and only if $\lim_{n \to \infty} \int_a^b f_n\, d\alpha$ is finite. Moreover $\int_a^b f\, d\alpha = \lim_{n \to \infty} \int_a^b f_n\, d\alpha$.*

The dominated convergence theorem need not be limited to increasing integrators. It costs nothing extra in the proof to deal with functions of bounded variation.

DOMINATED CONVERGENCE THEOREM. *Let $\int_a^b f_n \, d\alpha$ exist for $n = 1, 2, 3, \ldots$. Let $V_a^x \alpha$ exist whenever $[a, x]$ is contained in the domain of α. Suppose $\int_a^b h(x) \, dV_a^x \alpha$ exists and $|f_n| \leqslant h$ for all n. Finally suppose $\lim_{n \to \infty} f_n(x) = f(x)$ for all x in $[a, b]$. Then $\int_a^b f \, d\alpha$ exists and $\lim_{n \to \infty} \int_a^b f_n \, d\alpha = \int_a^b f \, d\alpha$.*

Note particularly that α need not be a real-valued function in this version of the dominated convergence theorem.

The proofs of these two theorems parallel the arguments given in Section S3.10. For the monotone convergence theorem the only change, apart from notational changes, is to replace the auxiliary function discussed in part (a) by a function such as the one in Example 5 above. The dominated convergence proof requires only one more change. That is to integrate all positive valued functions used in obtaining dominants with respect to the variation function of α.

7.9. Change of variables. There is a striking simplicity to change of variables in integrals of Stieltjes type. Suppose $\int_a^b f(x) \, d\alpha(x)$ is given. The change of variables $x = \tau(t)$ takes the form

$$\int_c^d f(\tau(t)) \, d\alpha(\tau(t)) = \int_{\tau(c)}^{\tau(d)} f(x) \, d\alpha(x).$$

Since it is not necessary to use the identity function as integrator, there is no need to require that τ be differentiable.

Generally speaking, the less f and α are restricted the more τ must be restricted. In all the following change-of-variables assertions τ is required to be continuous. Further restrictions on τ are designed to control oscillatory behavior to a greater or lesser degree.

The term "increasing" is used in the weak sense. That is, τ is increasing on $[c, d]$ when $\tau(t_1) \leqslant \tau(t_2)$ whenever $c \leqslant t_1 < t_2 \leqslant d$. A similar definition is to be understood for decreasing functions. A function is monotone on an interval if it is increasing there or decreasing there.

Let $\tau : [c, d] \to [a, b]$ be continuous and monotone. If $\int_a^b f \, d\alpha$ exists then $\int_c^d f \circ \tau \, d(\alpha \circ \tau)$ exists and equals $\int_{\tau(c)}^{\tau(d)} f \, d\alpha$.

The proof of this proposition deals directly with the definition. The major steps follow.

Suppose first that τ is increasing. If $\tau(c) = \tau(d)$, the function τ is constant and both integrals are zero. Thus we may suppose $\tau(c) < \tau(d)$.

The function τ maps intervals in $[c, d]$ onto intervals in $[\tau(c), \tau(d)]$. Even though some of its intervals may have degenerate images, each tagged division $\mathcal{D}$ of $[c, d]$ gives rise to a tagged division $\mathcal{D}'$ of $[\tau(c), \tau(d)]$. Moreover, $f \circ \tau \Delta(\alpha \circ \tau)(\mathcal{D}) = f \Delta \alpha(\mathcal{D}')$.

When a gauge γ' on $[\tau(c), \tau(d)]$ is given, the continuity of τ permits the formation of a gauge γ on $[c, d]$ so that $\mathcal{D}'$ is γ'-fine whenever $\mathcal{D}$ is γ-fine.

Since $\int_a^b f \, d\alpha$ is assumed to exist, it is also true that $\int_{\tau(c)}^{\tau(d)} f \, d\alpha$ exists. Thus there is a gauge γ' which insures that $f \Delta \alpha(\mathcal{D}')$ approximates $\int_{\tau(c)}^{\tau(d)} f \, d\alpha$ closely. When γ is determined from γ' the conclusion follows that $f \circ \tau \Delta(\alpha \circ \tau)(\mathcal{D})$ approximates $\int_{\tau(c)}^{\tau(d)} f \, d\alpha$ equally closely for any γ-fine $\mathcal{D}$. Thus, by definition, $\int_c^d f \circ \tau \, d(\alpha \circ \tau)$ exists and equals $\int_{\tau(c)}^{\tau(d)} f \, d\alpha$.

When τ is decreasing, the oriented integral must be used since $\tau(d) \leqslant \tau(c)$. Thus $\int_{\tau(c)}^{\tau(d)} f \, d\alpha$ is defined to be $-\int_{\tau(d)}^{\tau(c)} f \, d\alpha$. Moreover, $f \circ \tau \Delta(\alpha \circ \tau)(\mathcal{D}) = -f \Delta \alpha(\mathcal{D}')$ in this case. The main line of the argument may still be used.

The preceding proposition can be used as a tool to weaken the restrictions on τ somewhat without imposing any further restrictions on f and α.

We need some definitions. Say that τ is *piecewise monotone* on $[c, d]$ if there is a division of $[c, d]$ such that τ is monotone on each closed interval of the division. Define *piecewise one-to-one* similarly. Thus, for example, $\tau(t) = t^2$ is piecewise one-to-one on $[-1, 1]$, since it is one-to-one on each of $[-1, 0]$ and $[0, 1]$. When τ is continuous and piecewise monotone, it may have intervals of constancy. A continuous function which is piecewise one-to-one is also piecewise monotone but intervals of constancy are ruled out.

Let $\tau : [c, d] \to [a, b]$ be continuous and piecewise monotone. If $\int_a^b f \, d\alpha$ exists then $\int_c^d f \circ \tau \, d(\alpha \circ \tau)$ exists and equals $\int_{\tau(c)}^{\tau(d)} f \, d\alpha$. Under the added conditions that τ is piecewise one-to-one and maps $[c, d]$ onto $[a, b]$, there is the further conclusion that $\int_a^b f \, d\alpha$ exists only if $\int_c^d f \circ \tau \, d(\alpha \circ \tau)$ exists.

Suppose first that τ is piecewise monotone. The preceding result asserts that $\int_u^v f \circ \tau \, d(\alpha \circ \tau) = \int_{\tau(u)}^{\tau(v)} f \, d\alpha$ when $[u, v]$ is a subinterval of $[c, d]$ on which τ is monotone. Additivity of the integral assures us that $\int_c^d f \circ \tau \, d(\alpha \circ \tau) = \sum \int_u^v f \circ \tau \, d(\alpha \circ \tau)$ when the sum is taken over all $[u, v]$ in a division of $[c, d]$. The corresponding sum $\sum \int_{\tau(u)}^{\tau(v)} f \, d\alpha$ is not taken over a division of the interval with endpoints $\tau(c)$ and $\tau(d)$ since τ is not monotone on $[c, d]$. Nevertheless, an easy induction argument shows that it equals $\int_{\tau(c)}^{\tau(d)} f \, d\alpha$. (Keep in mind that the oriented integral is essential here.) Thus $\int_c^d f \circ \tau \, d(\alpha \circ \tau) = \int_{\tau(c)}^{\tau(d)} f \, d\alpha$.

Now impose the supplementary hypotheses on τ and assume $\int_c^d f \circ \tau \, d(\alpha \circ \tau)$ exists. We have to show that $\int_a^b f \, d\alpha$ exists. It suffices to find a division of $[a, b]$ such that f can be integrated with respect to α on each interval in the division.

Let τ be one-to-one on $[u, v]$. Then τ maps $[u, v]$ onto an interval $[r, s]$ contained in $[a, b]$. There is a continuous inverse $\sigma : [r, s] \to [u, v]$ such that $f = (f \circ \tau) \circ \sigma$ and $\alpha = (\alpha \circ \tau) \circ \sigma$ on $[r, s]$. Consequently, $\int_r^s f \, d\alpha$ exists because $\int_u^v (f \circ \tau) \, d(\alpha \circ \tau)$ exists. Now let $[u, v]$ range over the intervals in a division of $[c, d]$ chosen so that τ is one-to-one on each of them. The corresponding finite collection of intervals $[r, s]$ covers $[a, b]$ since τ maps $[c, d]$ onto $[a, b]$. However, they need not form a division of $[a, b]$ since nonoverlapping intervals in $[c, d]$ may have overlapping images. Any finite collection of intervals covering $[a, b]$ has a refinement which is a division of $[a, b]$. Since integrability on $[r, s]$ implies integrability on all subintervals of $[r, s]$, there is a division of $[a, b]$ such that f has an integral with respect to α on each interval of the division. Hence $\int_a^b f \, d\alpha$ exists.

The process of easing restrictions on τ by tightening restrictions on f and α can begin by using limit properties of the integral.

Exercise 8. Suppose $\int_a^b f \, d\alpha$ exists. Suppose $\tau : [c, d] \to [a, b]$ is continuous on $[c, d]$ and piecewise monotone on $[c, v]$ for all $v < d$. Suppose α is continuous at $\tau(d)$. Show that $\int_c^d f \circ \tau \, d(\alpha \circ \tau) = \int_{\tau(c)}^{\tau(d)} f \, d\alpha$.

Stringent restrictions on f and α allow for further easing of the conditions τ must meet.

Let f be continuous and α monotone on the bounded interval $[a, b]$. Let $\tau : [c, d] \to [a, b]$ be continuous and let $\alpha \circ \tau$ have bounded variation. Then $\int_c^d f \circ \tau \, d(\alpha \circ \tau) = \int_{\tau(c)}^{\tau(d)} f \, d\alpha$.

Both integrals are known to exist under these conditions on f, α, and τ. To show that they are equal express $\int_{\tau(c)}^{\tau(d)} f \, d\alpha - f \circ \tau \Delta(\alpha \circ \tau)(\mathcal{D})$ as $\sum \int_{\tau(u)}^{\tau(v)} [f(x) - f(\tau(z))] \, d\alpha(x)$. Here $\mathcal{D}$ is a division of $[c, d]$ and the sum is taken over all $z[u, v]$ in $\mathcal{D}$. The continuity of τ at z and f at $\tau(z)$ and the monotonicity of α enable us to claim that

$$\left| \int_{\tau(u)}^{\tau(v)} [f(x) - f(\tau(z))] \, d\alpha(x) \right| \leqslant \epsilon |\alpha(\tau(v)) - \alpha(\tau(u))|$$

when $[u, v]$ is contained in a small interval $\gamma(z)$. Thus, for γ-fine $\mathcal{D}$, $|\int_{\tau(c)}^{\tau(d)} f \, d\alpha - f \circ \tau \Delta(\alpha \circ \tau)(\mathcal{D})| \leqslant \epsilon V_c^d \alpha \circ \tau$. It follows that $\int_c^d f \circ \tau \, d(\alpha \circ \tau) = \int_{\tau(c)}^{\tau(d)} f \, d\alpha$.

Recall that we can convert integrals with respect to α into integrals with respect to the identity function when α is a primitive. This conversion opens the door to the development of other conditions under which the change of variables is valid. One instance is stated in Exercise 26 at the end of this chapter. It is, of course, directly related to some of the change of variables conditions given in Section 2.7.

7.10. Mean value theorems for integrals. The starting point is a mean value theorem for integrals of Stieltjes type.

Suppose f is continuous and real-valued on the bounded interval $[a, b]$. Let α be increasing on $[a, b]$. There exists c

in $[a, b]$ *such that*

$$\int_b^a f \, d\alpha = f(c)(\alpha(b) - \alpha(a)).$$

The function f has a minimum m and a maximum M on $[a, b]$. Since α is increasing,

$$m(\alpha(b) - \alpha(a)) \leqslant \int_a^b f \, d\alpha \leqslant M(\alpha(b) - \alpha(a)).$$

Let $g(t) = f(t)(\alpha(b) - \alpha(a))$. Then g is continuous with minimum $m(\alpha(b) - \alpha(a))$ and maximum $M(\alpha(b) - \alpha(a))$. Since g also has the intermediate value property, there is c in $[a, b]$ such that $g(c) = \int_a^b f \, d\alpha$, as claimed.

Note that it is not always possible to choose c so that $a < c < b$, as one can in the mean value theorem of differential calculus. For instance, the only possibility is $c = 1$ when $f(x) = x$, $\alpha(x) = 0$ for $0 \leqslant x < 1$, and $\alpha(1) = 1$.

This mean value theorem has as corollaries mean value theorems for integrals of the form $\int_a^b f g$. The derivation uses the conversion $\int_a^b f \, d\alpha = \int_a^b fg \, d\beta$ which holds when $\alpha(v) - \alpha(u) = \int_u^v g \, d\beta$ for all u and v. Of course we will use the identity function for β.

Before going into the mean value theorems it is proper to note some conditions under which $\int_a^b fg$ exists. First recall that products are not always integrable. For example, when $f(x) = g(x) = 1/\sqrt{x}$ on $(0, 1]$, $\int_0^1 fg$ does not exist. There are two sets of conditions on f and g which match the existence theorems for Riemann-Stieltjes integrals.

Suppose $\int_a^b g$ exists and f has bounded variation. Then $\int_a^b fg$ exists.

Now α is continuous when $\alpha(x) = \int_a^x g$. Since f has bounded variation $\int_a^b f \, d\alpha$ exists. So does $\int_a^b fg$.

Suppose $\int_a^b g$ and $\int_a^b |g|$ exist. Let f be a regulated function. Then $\int_a^b fg$ exists.

Since $|\alpha(v) - \alpha(u)| \leqslant \int_u^v |g|$, it follows that α is a function of bounded variation. Hence $\int_a^b f \, d\alpha$ exists for any regulated function f.

(The second of these is a special case of the result obtained in Section 5.4, p. 141, by more sophisticated methods.)

Suppose f is continuous and real-valued, $\int_a^b g$ exists, and $g \geqslant 0$. Then there is c in $[a, b]$ such that $\int_a^b fg = f(c)\int_a^b g$.

Since g is nonnegative, α is increasing. The result is immediate from the earlier mean value theorem.

Suppose f is increasing, $\int_a^b g$ exists, and g is real-valued. There exists c in $[a, b]$ such that

$$\int_a^b fg = f(a)\int_a^c g + f(b)\int_c^b g.$$

This time an integration by parts is needed so that the mean value theorem can be applied to $\int_a^b \alpha \, df$. A simple calculation gives the equation which has been asserted. (Keep in mind that $\alpha(a) = 0$.)

S7.11. Sequences of integrators. There are circumstances in which it is very helpful to know that $\lim_{n \to \infty} \int_a^b f \, d\alpha_n = \int_a^b f \, d\alpha$. (See Exercises 20–23 at the end of this chapter.) We should note at the outset that uniform convergence of α_n to α is not an appropriate condition.

Example 9. Let $f(0) = 0$ and $f(x) = x^{-2}$ when $x > 0$.

(a) Let $\alpha(x) = x$ for all x. Let $\alpha_n(x) = 0$ when $0 \leqslant x < 1/n$ and $\alpha_n(x) = x$ when $x \geqslant 1/n$. Then $\alpha_n(x)$ converges uniformly to $\alpha(x)$ on $[0, 1]$. We know that $\int_0^1 f \, d\alpha$ does not exist. Neither does $\lim_{n \to \infty} \int_0^1 f \, d\alpha_n$ since

$$\int_0^1 f \, d\alpha_n = \int_0^{1/n} f \, d\alpha_n + \int_{1/n}^1 f \, d\alpha_n = n + n - 1.$$

(b) Let $\alpha(x) = 0$ for all x. Let $\alpha_n(x) = 0$ when $0 \leqslant x < 1/n$ and $\alpha_n(x) = 1/n$ when $x \geqslant 1/n$. The convergence of $\alpha_n(x)$ to $\alpha(x)$ on $[0, 1]$ is again uniform. Moreover $\int_0^1 f \, d\alpha = 0$. But $\int_0^1 f \, d\alpha_n = n$.

Since $\lim_{n \to \infty} \int_a^b f \, d\alpha_n = \lim_{n \to \infty} (\lim_{\mathfrak{D}} f \Delta \alpha_n(\mathfrak{D}))$, we are dealing with an iterated limits problem. It is natural to seek conditions under which one of the inner limits is uniform. The following statement includes one of each type.

Suppose that $\int_a^b f \, d\alpha_n$ exists for all n. Then $\int_a^b f \, d\alpha$ and $\lim_{n \to \infty} \int_a^b f \, d\alpha_n$ exist and are equal under each of the following additional conditions.

   (i) $V_a^b(\alpha_n - \alpha) \to 0$ as $n \to \infty$ and f is bounded.

   (ii) $V_a^b \alpha_n$ is bounded, f is continuous, and $\lim_{n \to \infty} \alpha_n = \alpha$.

Under condition (i) $f \Delta \alpha_n(\mathfrak{D})$ converges uniformly to $f \Delta \alpha(\mathfrak{D})$ for all tagged divisions $\mathfrak{D}$. This is clear from the fact that

$$|f \Delta \alpha_n(\mathfrak{D}) - f \Delta \alpha(\mathfrak{D})|$$
$$= |f \Delta(\alpha_n - \alpha)(\mathfrak{D})| \leqslant |f||\Delta(\alpha_n - \alpha)|(\mathfrak{D})$$

for every tagged division $\mathfrak{D}$.

When (ii) holds, $f\Delta\alpha_n(\mathcal{D})$ converges to $\int_a^b f\,d\alpha_n$ uniformly for all n. This is most easily established by observing that uniform continuity of f implies a uniform Cauchy criterion for existence of $\int_a^b f\,d\alpha_n$. We are able to use the norm integral here since f is continuous and each α_n is a function of bounded variation.

Choose δ so that $|f(x) - f(y)| < \epsilon$ when $|x - y| \leqslant \delta$. Let M be a bound for $V_a^b\alpha_n$. Suppose $\|\mathcal{D}\| \leqslant \delta$ and $\|\mathcal{E}\| \leqslant \delta$. Let $\mathcal{F}$ be a common refinement of $\mathcal{D}$ and $\mathcal{E}$ with any tags on $\mathcal{F}$ one cares to use. It suffices to compare $f\Delta\alpha_n(\mathcal{F})$ to each of $f\Delta\alpha_n(\mathcal{D})$ and $f\Delta\alpha_n(\mathcal{E})$. The conclusion is that $f\Delta\alpha_n(\mathcal{F})$ is within a distance $M\epsilon$ of each and thus $|f\Delta\alpha_n(\mathcal{D}) - f\Delta\alpha_n(\mathcal{E})| < 2M\epsilon$. Consequently, by taking the limit on $\mathcal{E}$, $|f\Delta\alpha_n(\mathcal{D}) - \int_a^b f\,d\alpha_n| \leqslant 2M\epsilon$ for every n and every $\mathcal{D}$ for which $\|\mathcal{D}\| \leqslant \delta$. Moreover, the convergence of α_n to α implies that $\lim_{n\to\infty} f\Delta\alpha_n(\mathcal{D}) = f\Delta\alpha(\mathcal{D})$ for every $\mathcal{D}$.

Now the iterated limits theorem of Section S3.9 can be applied under each of the conditions (i) and (ii).

In Example 9, parts (a) and (b), $V_0^1(\alpha_n - \alpha)$ has a zero limit. Thus the boundedness of f cannot be dispensed with in condition (i). The next example shows that $V_a^b(\alpha_n - \alpha) \to 0$ cannot be weakened to boundedness of $V_a^b\alpha_n$ in condition (i).

Example 10. Let $f(x) = \alpha(x) = 0$ when $0 \leqslant x < 1$ and $f(x) = \alpha(x) = 1$ when $1 \leqslant x \leqslant 2$. Let $\alpha_n(x) = 0$ when $0 \leqslant x \leqslant 1 - 1/n$ and $\alpha_n(x) = 1$ when $1 \leqslant x \leqslant 2$. Let α_n rise linearly from 0 to 1 on $[1 - 1/n, 1]$. Then $\lim_{n\to\infty}\alpha_n(x) = \alpha(x)$ on $[0, 2]$. Moreover $V_0^2\alpha_n = 1$ for all n. Now $\int_0^2 f\,d\alpha_n = 0$ for all n but $\int_0^2 f\,d\alpha = 1$. Thus $\int_0^2 f\,d\alpha \neq \lim_{n\to\infty}\int_0^2 f\,d\alpha_n$.

This last example shows equally well that in condition (ii) it is not possible to replace continuity of f by boundedness of f. Neither can the bound on the variations be removed while continuity of f is retained.

Exercise 9. Construct an example in which f is continuous, $V_a^b \alpha_n$ is not bounded, α_n converges uniformly to α, and $\lim_{n \to \infty} \int_a^b f \, d\alpha_n = \int_a^b f \, d\alpha$ is not true.

S7.12. Line integrals. Line integrals are integrals of Stieltjes type associated with parametrically defined curves.

Let $\alpha : [a, b] \to \mathbf{R}^p$ be a given continuous function. Let E be the set of values of α. Let f and g be functions defined at least on E. Suppose the values of f and g can be multiplied. The integral of f with respect to g over α is defined by

$$\int_\alpha f \, dg = \int_a^b f \circ \alpha \, d(g \circ \alpha)$$

whenever the integral on the right is meaningful.

There are various sets of additional restrictions on f, g, and α which insure existence of $\int_\alpha f \, dg$. Here is one which is commonly met. Let f be continuous on E. Let g be Lipschitz continuous on E. That is, suppose there is a constant M such that $|g(x) - g(y)| \leq M |x - y|$ for all x and y in E. Let α be a function of bounded variation on $[a, b]$. It is immediate that $g \circ \alpha$ also has bounded variation. Consequently, the continuous function $f \circ \alpha$ is integrable with respect to $g \circ \alpha$.

The geometrical significance of bounded variation for α is that α defines a curve of finite length. This was pointed out in Section 3.2.

The function α may be thought of as describing the motion of a particle. The position of the particle at time t

is $\alpha(t)$. As t moves from a to b the particle traces E in some manner from $\alpha(a)$ to $\alpha(b)$. The motion may not always go steadily forward. The particle may hesitate, rest, and retrace portions of its journey in the same or opposite sense or both.

Let $\beta = \alpha \circ \tau$ where τ is a continuous and monotone function from $[c, d]$ onto $[a, b]$. Then β also traces over the same set E as α. It moves in the same direction as α when τ is increasing and in the opposite sense when τ is decreasing. The first change of variable theorem (p. 206) says that

$$\int_c^d (f \circ \alpha) \circ \tau \, d((g \circ \alpha) \circ \tau) = \int_{\tau(c)}^{\tau(d)} f \circ \alpha \, d(g \circ \alpha).$$

Since function composition is associative the integral on the left is $\int_\beta f \, dg$. When τ is increasing the right-hand side is $\int_\alpha f \, dg$ because $\tau(c) = a$ and $\tau(d) = b$. When τ is decreasing the right-hand side is $-\int_\alpha f \, dg$ because $\tau(c) = b$ and $\tau(d) = a$.

When τ is increasing, α and β can be regarded as equivalent. When τ is decreasing, they are opposites or negatives. It is common to write $\beta = -\alpha$ in this case. This must not be confused with the entirely different negative obtained by multiplying function values by -1.

Example 11. Let $\alpha(t) = (\sin t, \cos t)$, $0 \leqslant t \leqslant \pi/2$. Let $\beta(t) = (-\sin t, \cos t)$, $-\pi/2 \leqslant t \leqslant 0$. Clearly $\beta = \alpha \circ \tau$ where $\tau(t) = -t$ when $-\pi/2 \leqslant t \leqslant 0$. The set E is the first quadrant portion of the circle $x^2 + y^2 = 1$. It is traced from $(0, 1)$ to $(1, 0)$ by α and in the reverse sense by β.

It often happens that a function α occurs in a very natural way but its opposite is the one which is needed. The introduction of a decreasing change of parameter τ is not difficult but it becomes a nuisance if it is needed frequently. Moreover, it is really unnecessary. The same effect on the tracing of the image set E and on line integrals can be achieved by keeping the same interval and function but reversing the orientation of the interval on which α is defined. For instance, let $\alpha(t) = (t, t)$ when t is between 0 and 2. Let [2, 0] be the numbers between 0 and 2 ordered by $\geqslant$. That is, 2 is the initial point and 0 is the terminal point. Consequently, the tracing point $\alpha(t)$ moves from $\alpha(2)$ to $\alpha(0)$. Moreover,

$$\int_\alpha f \, dg = \int_2^0 f \circ \alpha \, d(g \circ \alpha).$$

In general, suppose we are given a function α on an oriented interval $[a, b]$ where a is the initial point and b is the terminal point. Then $\int_\alpha f \, dg$ is defined as before except that the defining integral is an oriented integral. To call attention to the orientation, let's call α defined on an oriented interval a *path*.

No exhaustive discussion of integrals on paths is intended. We shall only indicate some uses of the foregoing treatment of integrals by considering Green's Theorem for plane regions.

Let f be defined on a set E in $\mathbf{R}^2$. Let $g(x, y) = x$ for all (x, y) in $\mathbf{R}^2$. Then $\int_\alpha f \, dg$ is usually written $\int_\alpha f(x, y) \, dx$. A similar meaning is attached to $\int_\alpha f(x, y) \, dy$. Also, $\int_\alpha P(x, y) \, dx + Q(x, y) \, dy$ is, by definition, $\int_\alpha P(x, y) \, dx + \int_\alpha Q(x, y) \, dy$.

A *chain* is a finite family of paths. The line integral over a chain is the sum of the integrals over the individual paths.

Let α be a chain which traces the boundary of a plane region R. Green's Theorem says that, under suitable conditions,

$$\int_\alpha P(x, y)\, dx + Q(x, y)\, dy = \int_R [Q_1 - P_2]$$

where P_2 is the partial derivative of $P(x, y)$ with respect to y and Q_1 is the partial derivative of $Q(x, y)$ with respect to x.

The terms involving P can be handled separately from those involving Q. At first, conditions will be placed on R suited to the calculations with the P terms.

Let ϕ_1 and ϕ_2 be continuous functions such that $\phi_1(x) \leqslant \phi_2(x)$ for $a \leqslant x \leqslant b$. Let R be the region between the graphs of $y = \phi_1(x)$ and $y = \phi_2(x)$. Then the chain α made up of the following paths traces the boundary of R counterclockwise:

$$\alpha_1(t) = (t, \phi_1(t)), \qquad a \leqslant t \leqslant b;$$

$$\alpha_2(t) = (b, t), \qquad \phi_1(b) \leqslant t \leqslant \phi_2(b);$$

$$\alpha_3(t) = (t, \phi_2(t)), \qquad b \geqslant t \geqslant a;$$

$$\alpha_4(t) = (a, t), \qquad \phi_2(a) \geqslant t \geqslant \phi_1(a).$$

One or both of α_2 and α_4 may degenerate to a single point. Any degenerate paths may be left out. The paths are shown in Figure 1(a) with the orientations marked with arrows. Since x is constant on α_2 and α_4,

$$\int_\alpha P(x, y)\, dx = \int_a^b P(x, \phi_1(x))\, dx + \int_b^a P(x, \phi_2(x))\, dx$$

$$= -\int_a^b [P(x, \phi_2(x)) - P(x, \phi_1(x))]\, dx.$$

For this it suffices that P be continuous on the graphs of

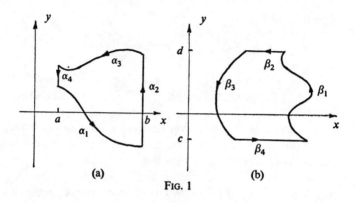

(a)   (b)

FIG. 1

ϕ_1 and ϕ_2. The Fundamental Theorem asserts that

$$P(x, v) - P(x, u) = \int_u^v P_2(x, y) \, dy$$

when $P(x, y)$ is continuous in y and $P_2(x, y)$ exists except possibly for a countable set of y's such that $u \leqslant y \leqslant v$.

Let C be a countable set of interior points of R. Suppose P is continuous on $R - C$ and P_2 exists on $R - C$. Let P_2 be integrable on R. That is, the function which is P_2 on R and zero outside is integrable on a rectangle S containing R. Apply Fubini's Theorem to $-P_2$ on S. Let the inner integration be with respect to y. Then the Fundamental Theorem can be applied for all but a countable set of values of y. The result is

$$-\int_R P_2 = \int_\alpha P(x, y) \, dx.$$

Next suppose R is the region between $x = \psi_1(y)$ and $x = \psi_2(y)$ where $\psi_1(y) \leqslant \psi_2(y)$ for $c \leqslant y \leqslant d$. Let β be the chain tracing the boundary of R counterclockwise as shown in Figure 1(b). Now

$$\int_{\beta} Q(x,y)\,dy = \int_{c}^{d} \left[Q(\psi_2(y), y) - Q(\psi_1(y), y) \right]\,dy.$$

Let Q and Q_1 satisfy conditions like those imposed on P and P_2. Then

$$\int_{R} Q_1 = \int_{\beta} Q(x,y)\,dy.$$

Now let's take a region R which is simultaneously of the two types shown in Figure 1. It might look like the region shown in Figure 2.

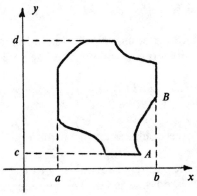

Fig. 2

When P and Q satisfy the conditions above,

$$\int_{\alpha} P(x,y)\,dx + \int_{\beta} Q(x,y)\,dy = \int_{R} \left[Q_1 - P_2 \right].$$

The path integrals can both be taken over α. This can be seen by breaking α and β down into subpaths in a natural way. For instance, the arc from A to B in Figure 2 is traced by a part of α_1 and a part of β_1. It is easy to see that a change of parameter links the two.

The foregoing may be summarized as follows.

GREEN'S THEOREM.   *Let R be a region which is of both types shown in Figure 1. Let P and Q be functions defined on R. Suppose there is a countable set C of interior points of R such that the partial derivatives P_2 and Q_1 exist on $R - C$. Let P and Q be continuous on $R - C$ and let P_2 and Q_1 be integrable on R. Let α be a chain tracing the boundary of R once counterclockwise. Then*

$$\int_\alpha P(x, y)\, dx + Q(x, y)\, dy = \int_R \left[Q_1 - P_2 \right].$$

It is definitely not enough that the difference $Q_1 - P_2$, rather than the terms Q_1 and P_2, be integrable.

Example 12.   Let $P(x, y) = -y/(x^2 + y^2)$ and $Q(x, y) = x/(x^2 + y^2)$. Let R be the disk where $x^2 + y^2 \leqslant 1$. Now $Q_1(x, y) = P_2(x, y) = (y^2 - x^2)/(x^2 + y^2)^2$ except at $(0, 0)$. Thus the double integral on the right is zero. The path $\alpha_1(t) = (\cos t, \sin t)$, $0 \leqslant t \leqslant 2\pi$, gives the same integrals as the natural chain α. Using α_1 it is easy to show that the left-hand side is 2π. The nonintegrability of Q_1 and P_2 has already been discussed in Example 6 of Section 1.6.

S7.13.   Functions of bounded variation and regulated functions.   This discussion is limited to those facts which proved indispensable in earlier parts of the chapter. It is a self-contained treatment of them.

Recall that $f : [a, b] \to \mathbf{R}^q$ is a function of bounded variation when $\text{lub}_\mathscr{D} |\Delta f|(\mathscr{D})$ is finite and that the total variation, $V_a^b f$, is defined by $V_a^b f = \text{lub}_{\mathcal{D}} |\Delta f|(\mathcal{D})$. One of the most useful properties of the total variation is its additivity. We shall show that *bounded variation on* $[a, b]$ *implies bounded variation on each subinterval of* $[a, b]$ and, conversely, that *bounded variation on* $[a, c]$ *and on* $[c, b]$ *implies bounded variation on* $[a, b]$. Moreover, $V_a^b f = V_a^c f + V_c^b f$.

Let $[c, d] \subseteq [a, b]$. Let $\mathcal{D}$ be a division of $[c, d]$. Adjoin $[a, c]$ and $[d, b]$ to $\mathcal{D}$ to form $\mathcal{E}$. Now $\mathcal{E}$ is a division of $[a, b]$ and

$$|\Delta f|(\mathcal{E}) = |\Delta f|(\mathcal{D}) + |f(c) - f(a)| + |f(b) - f(d)|.$$

Consequently

$$|\Delta f|(\mathcal{D}) \leqslant V_a^b f - (|f(c) - f(a)| + |f(b) - f(d)|).$$

It follows that f has bounded variation on $[c, d]$.

Now assume $V_a^c f$ and $V_c^b f$ exist with $a < c < b$. Let $\mathcal{D}$ be a division of $[a, b]$. Either c is an endpoint of adjacent intervals of $\mathcal{D}$ or it lies within an interval of $\mathcal{D}$. Either way there are $\mathcal{E}$ and $\mathcal{F}$ so that $\mathcal{E} \cup \mathcal{F}$ is a refinement of $\mathcal{D}$ and individually they are divisions of $[a, c]$ and $[c, b]$. Then

$$|\Delta f|(\mathcal{D}) \leqslant |\Delta f|(\mathcal{E}) + |\Delta f|(\mathcal{F}) \leqslant V_a^c f + V_c^b f.$$

It follows that $V_a^b f \leqslant V_a^c f + V_c^b f$.

On the other hand, given $\mathcal{E}$ and $\mathcal{F}$ which are divisions of $[a, c]$ and $[c, b]$, set $\mathcal{D} = \mathcal{E} \cup \mathcal{F}$. Then $\mathcal{D}$ is a division of $[a, b]$ and $|\Delta f|(\mathcal{D}) = |\Delta f|(\mathcal{E}) + |\Delta f|(\mathcal{F})$. For well-chosen $\mathcal{E}$ and $\mathcal{F}$,

$$V_a^c f + V_c^b f - \epsilon < |\Delta f|(\mathcal{D}) \leqslant V_a^b f.$$

Consequently, $V_a^c f + V_c^b f \leqslant V_a^b f$. This completes the proof of additivity.

When $V_a^b f$ exists the rule $x \to V_a^x f$ defines the *variation function* of f on $[a, b]$. (It is agreed that $V_a^a f = 0$.) Since $V_x^y f \geqslant 0$ and $V_a^y f = V_a^x f + V_x^y f$ when $x < y$, the variation function is an increasing function.

For any increasing function g it is easy to establish the existence of a left-hand limit $g(x-)$ and a right-hand limit $g(x+)$. Indeed they are $\text{lub}\{ g(y) : y < x \}$ and $\text{glb}\{ g(y) : x < y \}$, respectively. Obviously $g(x-) \leqslant g(x) \leqslant g(x+)$ but equality need not hold everywhere.

The existence of one-sided limits for a function of bounded variation can be obtained from the one-sided limits of the variation function through the use of the Cauchy criterion. Fix z with $a < z$. The existence of the left-hand limit of $V_a^x f$ at z implies that $|V_a^y f - V_a^x f| < \epsilon$ when x and y are in $(z - \delta, z)$ for a suitable δ. For convenience suppose $z - \delta < x < y < z$. Then $V_a^y f - V_a^x f = V_x^y f$. Moreover, $|f(y) - f(x)| \leqslant V_x^y f$. Thus $|f(y) - f(x)| < \epsilon$ when x and y are in $(z - \delta, z)$. This is the Cauchy criterion for existence of $f(z-)$. A similar argument shows that the right-hand limit $f(z+)$ exists also.

A function which has one-sided limits everywhere is a *regulated* function. Thus *every function of bounded variation is a regulated function.*

The differences $f(z) - f(z-)$ and $f(z) - f(z+)$ show how much f misses being continuous at z. When f has bounded variation these expressions are severely limited in magnitude.

Suppose $f : [a, b] \to \mathbf{R}^q$ is a function of bounded variation. Let $c_1, c_2, c_3, \ldots$ be distinct points in (a, b). Then $\sum_{n=1}^{\infty} |f(c_n) - f(c_n -)|$ and $\sum_{n=1}^{\infty} |f(c_n) - f(c_n +)|$ are convergent.

Fix n for the moment. Select $d_1, d_2, \ldots, d_n$ so that $[c_1, d_1], [c_2, d_2], \ldots, [c_n, d_n]$ are nonoverlapping intervals in $[a, b]$ with $|f(c_i +) - f(d_i)| < 2^{-i}$ for $i = 1, 2, \ldots, n$. Then

$$\sum_{i=1}^{n} |f(c_i) - f(c_i +)|$$

$$\leqslant \sum_{i=1}^{n} |f(c_i) - f(d_i)| + \sum_{i=1}^{n} |f(d_i) - f(c_i +)|$$

$$\leqslant V_a^b f + 1.$$

Thus the partial sums of this series of nonnegative terms are bounded; hence the series is convergent. The series containing the left-hand limit may be handled similarly.

Next let us *suppose that* $f : [a, b] \rightarrow \mathbf{R}^q$ *has bounded variation on each* $[a, t]$ *with* $t < b$. *Then* f *has bounded variation on* $[a, b]$ *if and only if* $\lim_{t \to b} V_a^t f$ *exists. Moreover*

$$V_a^b f = \lim_{t \to b} V_a^t f + |f(b -) - f(b)|.$$

Suppose $\lim_{t \to b} V_a^t f$ exists. Call it A for convenience. The argument for existence of left-hand limits of functions of bounded variation given above did not use existence of $V_a^b f$ but only the left-hand limit of $V_a^t f$. Thus it is valid here. Consequently $f(b -)$ exists and $\lim_{t \to b} |f(t) - f(b)| = |f(b -) - f(b)|$. Call this number B for convenience.

Select δ so that $\| |f(t) - f(b)| - B| < \epsilon$ and $|V_a^t f - A| < \epsilon$ when $b - \delta < t < b$.

Let $\mathcal{D}$ be any division of $[a, b]$. We need to find a bound for $|\Delta f|(\mathcal{D})$. We may form a refinement $\mathcal{D}'$ containing an interval $[t, b]$ with $b - \delta < t < b$. Then

$$|\Delta f|(\mathcal{D}) \leqslant |\Delta f|(\mathcal{D}') \leqslant V_a^t f + |f(t) - f(b)|$$

$$\leqslant A + B + \epsilon.$$

Consequently $V_a^b f \leqslant A + B$.

On the other hand, when $b - \delta < t < b$,

$$A + B - 2\epsilon < V_a^t f + |f(t) - f(b)| \leqslant V_a^t f + V_t^b f = V_a^b f.$$

Thus $A + B \leqslant V_a^b f$. Equality follows from the two inequalities.

We already know that $\lim_{t \to b} V_a^t f$ exists when $V_a^b f$ exists. Thus the proof is complete.

We have seen earlier that every function of bounded variation is a regulated function, i.e., a function which has one-sided limits at each point. Now let's go on to a basic proposition about regulated functions.

A function $f : [a, b] \to \mathbf{R}^q$ is a regulated function if and only if f is the uniform limit of a sequence of step functions.

Suppose f is a regulated function. Let a positive number ϵ be given. We shall show that there is a step function g such that $|f(x) - g(x)| < \epsilon$ for every x in $[a, b]$.

Suppose $a \leqslant z < b$. The existence of $f(z +)$ implies the existence of a positive number δ such that $|f(x) - f(y)| < \epsilon$ when x and y are in $(z, z + \delta)$. When $a < z \leqslant b$ a similar argument yields δ so that $|f(x) - f(y)| < \epsilon$ when x and y are in $(z - \delta, z)$. When $a < z < b$, we may use the same δ on both sides by taking the smaller of the two. (Of course δ depends on z.)

Set $\gamma(z) = (z - \delta, z + \delta)$ for all z in $[a, b]$. Let $\mathcal{D}$ be a γ-fine division of $[a, b]$. We define the step function g from the intervals of $\mathcal{D}$. Let $z[u, v] \in \mathcal{D}$. Set $g(u) = f(u)$, $g(z) = f(z)$, and $g(v) = f(v)$. When $x \in (u, z)$ let $g(x) = f((u + z)/2)$. When $x \in (z, v)$ let $g(x) = f((z + v)/2)$. Since $[u, v] \subseteq \gamma(z)$ the choice of $\gamma(z)$ makes it clear that $|f(x) - g(x)| < \epsilon$ for all x in $[u, v]$. As $z[u, v]$ ranges over $\mathcal{D}$ this procedure defines g as a step function on $[a, b]$. Of course $|f(x) - g(x)| < \epsilon$ for all x in $[a, b]$.

Now take $\epsilon = 1/n$ for $n = 1, 2, 3, \ldots$ and form corresponding step functions f_n. Consequently, f is the uniform limit of $(f_n)_{n=1}^{\infty}$.

Suppose, conversely, that f is the uniform limit of some sequence of step functions $(f_n)_{n=1}^{\infty}$. The iterated limits theorem of Section S3.9 will enable us to conclude that f is a regulated function.

Since $\lim_{n \to \infty} f_n(x) = f(x)$ is uniform in x, it is permissible to interchange a limit on x with the limit on n. Thus

$$\lim_{x \to z} f(x) = \lim_{x \to z} \left(\lim_{n \to \infty} f_n(x) \right) = \lim_{n \to \infty} \left(\lim_{x \to z} f_n(x) \right)$$

when x approaches z from one side or the other provided $\lim_{x \to z} f_n(x)$ exists. But each f_n has one-sided limits everywhere. Thus f enjoys the same property.

We can draw an added conclusion from this interchange of limits. Whenever each f_n is continuous at z, the same is true of f. Each f_n has a finite set of discontinuities. The union of these sets of discontinuities is a countable set. Consequently *the set of discontinuities of a regulated function is a countable set.*

S7.14. Proof of the absolute integrability theorem.

On page 202 the absolute integrability theorem was put as follows. *Suppose $\int_a^b f \, d\alpha$ exists and α is a function of bounded variation on $[a, b]$. Set $F(x) = \int_a^x f \, d\alpha$ when $a \leqslant x \leqslant b$. Then $V_a^b F$ exists if and only if $\int_a^b |f(x)| \, dV_a^x \alpha$ exists. Moreover $V_a^b F = \int_a^b |f(x)| \, dV_a^x \alpha$.*

For convenience set $\beta(x) = V_a^x \alpha$.

One implication is easy to prove. Suppose $\int_a^b |f| \, d\beta$ exists. Since $|\Delta \alpha| \leqslant \Delta \beta$ we know that $|\int_u^v f \, d\alpha| \leqslant \int_u^v |f| \, d\beta$ for every $[u, v]$. Hence $|\Delta F|(\mathcal{D}) \leqslant \int_a^b |f| \, d\beta$ for every division $\mathcal{D}$ of $[a, b]$. Thus $V_a^b F$ exists and does not exceed $\int_a^b |f| \, d\beta$.

The more challenging part begins with the assumption that F is a function of bounded variation on $[a, b]$. We shall show directly from the definition of the integral that $V_a^b F$ is the integral of $|f|$ with respect to β.

We are obliged to approximate $V_a^b F$ by $|f| \Delta \beta (\mathcal{D})$ for γ-fine $\mathcal{D}$. This goes by way of a chain of approximations. The natural links from $V_a^b F$ to $|f| \Delta \beta (\mathcal{D})$ are $|\Delta F|(\mathcal{D})$ and $|f||\Delta \alpha|(\mathcal{D})$. This must be modified, however, because two kinds of conditions must be imposed to make the approximations good ones. Refinement enables us to get $|\Delta F|(\mathcal{D})$ close to $V_a^b F$. Use of a gauge brings $|f||\Delta \alpha|(\mathcal{D})$ near $|\Delta F|(\mathcal{D})$ through Henstock's lemma. A combination of refinement and use of a gauge draws $|f||\Delta \alpha|(\mathcal{D})$ and $|f| \Delta \beta (\mathcal{D})$ close to one another.

The modification is this. With each $\mathcal{D}$ we associate a companion $\mathcal{E}$ by breaking $z[u, v]$ of $\mathcal{D}$ into $z[u, z]$ and $z[z, v]$ of $\mathcal{E}$ whenever $u < z < v$. When z is already an endpoint, $z[u, v]$ goes into $\mathcal{E}$ unaltered. Then the chain from $V_a^b F$ to $|f| \Delta \beta (\mathcal{D})$ has $|\Delta F|(\mathcal{E})$, $|f||\Delta \alpha|(\mathcal{E})$, and $|f| \Delta \beta (\mathcal{E})$ as its links.

Begin by fixing a division $\mathcal{K}$ of $[a, b]$ so that $V_a^b F - \epsilon < |\Delta F|(\mathcal{D}) \leqslant V_a^b F$ whenever $\mathcal{D}$ is a refinement of $\mathcal{K}$. Also select a gauge γ so that $|\int_a^b f \, d\alpha - f \Delta \alpha (\mathcal{D})| < \epsilon$ when $\mathcal{D}$ is γ-fine. It is no loss to suppose that $\gamma(z)$ contains no endpoint of $\mathcal{K}$ distinct from z.

For each positive integer n fix a division $\mathcal{F}_n$ so that $V_a^b \alpha - \epsilon/(n2^n) < |\Delta \alpha|(\mathcal{D}) \leqslant V_a^b \alpha$ when $\mathcal{D}$ is a refinement of $\mathcal{F}_n$. Also let E_n be the set of all x in $[a, b]$ where $n - 1 \leqslant |f(x)| < n$. For each z in E_n restrict $\gamma(z)$ further so that $\gamma(z)$ contains no endpoint of $\mathcal{F}_n$ distinct from z.

Now we shall check each link in the chain of approximations. We begin with a tagged division $\mathcal{D}$ which is γ-fine and form $\mathcal{E}$ from $\mathcal{D}$.

The division $\mathcal{E}$ is a refinement of $\mathcal{K}$ since $\gamma(z)$ contains no endpoint of $\mathcal{K}$ distinct from z. Thus $|V_a^b F - |\Delta F|(\mathcal{E})| < \epsilon$.

Since $\|\Delta F|(\mathcal{E}) - |f|\|\Delta \alpha|(\mathcal{E})| \leqslant |\Delta F - f\Delta \alpha|(\mathcal{E})$, Henstock's lemma gives the estimate $2q\epsilon$ here since $\mathcal{E}$ is also γ-fine. (As before, q is the dimension of the space in which $f\alpha$ has its values.)

Now we must determine how close $|f|\|\Delta \alpha|(\mathcal{E})$ is to $|f|\Delta \beta(\mathcal{E})$. Let $\mathcal{E}_n$ be the subset of $\mathcal{E}$ consisting of all intervals with tags in E_n. These subsets form a partition of $\mathcal{E}$ since each point of $[a, b]$ falls into one and only one set E_n. Each interval of $\mathcal{E}_n$ is contained in an interval of $\mathcal{F}_n$. This follows from the restriction on γ in relation to $\mathcal{F}_n$. Then $0 \leqslant (\Delta \beta - |\Delta \alpha|)(\mathcal{E}_n) < \epsilon/(n2^n)$ because this sum can be enlarged by addition of nonnegative terms to become the sum for a division of $[a, b]$ which is a refinement of $\mathcal{F}_n$. Thus

$$0 \leqslant |f|\Delta \beta(\mathcal{E}) - |f|\|\Delta \alpha|(\mathcal{E}) = |f|(\Delta \beta - |\Delta \alpha|)(\mathcal{E})$$

$$= \sum_n |f|(\Delta \beta - |\Delta \alpha|)(\mathcal{E}_n) \leqslant \sum_n n(\Delta \beta - |\Delta \alpha|)(\mathcal{E}_n)$$

$$< \sum_n n(\epsilon/(n2^n)) = \epsilon.$$

Lastly, note that $|f|\Delta \beta(\mathcal{E}) = |f|\Delta \beta(\mathcal{D})$ because of the way $\mathcal{E}$ is formed.

From the triangle inequality, $|V_a^b F - |f|\Delta \beta(\mathcal{D})| < \epsilon + 2q\epsilon + \epsilon$. This completes the argument.

7.15. Exercises.

10. Consider the dot product in $\mathbf{R}^q$ and the vector cross product in $\mathbf{R}^3$. Define $\int_a^b f \cdot d\alpha$ and $\int_a^b f \times d\alpha$ as the gauge limits of sums formed from terms $f(z) \cdot \Delta \alpha(J)$ and $f(z) \times \Delta \alpha(J)$, respectively. Confirm linearity in integrator and integrand and additivity for these integrals.

11. Suppose f and α are both defined on $[a, b]$.

(a) Suppose f and α are both discontinuous at some point c in $[a, b]$. Show that $(\mathfrak{N})\int_a^b f\, d\alpha$ cannot exist.

(b) Suppose f and α are both discontinuous on the same side at some point c in $[a, b]$. Show that $(\mathfrak{R})\int_a^b f\, d\alpha$ cannot exist.

(c) Suppose $f(c) \neq 0$ and α fails to have a limit on one side at c. Show that $(\mathcal{G})\int_a^b f\, d\alpha$ cannot exist.

12. Let f be a regulated function on $[a, b]$. Set $F_0(x) = f(x) - f(a)$. Let F_n be a primitive of F_{n-1} with $F_n(a) = 0$ for $n = 1, 2, 3, \ldots$. Show that

$$\int_a^b x^n\, df(x) = b^n F_0(b) - n b^{n-1} F_1(b) + n(n-1) b^{n-2} F_2(b)$$

$$- \cdots + (-1)^n n!\, F_n(b).$$

13. Suppose $\sum_{n=1}^\infty |t_n| < \infty$. Let $(c_n)_{n=1}^\infty$ be a sequence in $[a, b]$ such that $c_m \neq c_n$ when $m \neq n$.

(a) Set $f_n(x) = 1$ when $a \leqslant x \leqslant c_n$ and $f_n(x) = 0$ when $c_n < x$. Show that $\sum_{n=1}^\infty t_n f_n$ converges uniformly on $[a, b]$ to a function α of bounded variation. Evaluate $V_a^b \alpha$. Determine $\alpha(x+) - \alpha(x)$ and $\alpha(x-) - \alpha(x)$.

(b) Similar requests when $f_n(x) = 1$ for $a \leqslant x < c_n$ and $f_n(x) = 0$ for $c_n \leqslant x$.

14. Suppose f is a function of bounded variation on $[a, b]$. Use Exercise 13 to define a function f_s of bounded variation such that $f - f_s$ is continuous.

15. Suppose $\sum_{n=1}^\infty |t_n| < \infty$. Let $\epsilon_n = \pm 1$ for all n. Construct a function f on $[0, 1]$ such that

$$\int_0^1 f\, df = \frac{1}{2}\left[\sum_{n=1}^\infty t_n \right]^2 + \frac{1}{2} \sum_{n=1}^\infty \epsilon_n t_n^2.$$

16. Let f be regulated on $[a, b]$. Let $(c_n)_{n=1}^\infty$ be a sequence of distinct points of $[a, b]$. Is $\sum_{n=1}^\infty |f(c_n) - f(c_n+)|$ finite?

17. Let $f(x) = x^{-3/2} \sin 1/x$ when $x > 0$. Show $\int_0^1 f$ exists but $\int_0^1 |f|$ does not exist. Find a continuous function g such that $\int_0^1 fg$ does not exist.

18. Use the version of Henstock's lemma which applies to the total variation to show that

$$\lim_{x \to z+} V_z^x \alpha = |\alpha(z) - \alpha(z+)|,$$

$$\lim_{x \to z-} V_x^z \alpha = |\alpha(z) - \alpha(z-)|.$$

Draw conclusions about the continuity of the variation function of α.

19. Let $\alpha(1/n) = (-1)^n/(n^2 + n)$ for $n = 1, 2, 3, \ldots$. Let $\alpha(x) = 0$ elsewhere in $[0, 2]$.

 (a) Show that $V_0^2 \alpha = 2$.

 (b) Find a nonnegative function f on $[0, 2]$ such that $\int_0^2 f \, d\alpha$ exists but $\int_0^2 f(x) \, dV_0^x \alpha$ does not.

The next four exercises are interrelated.

20. Let $\lambda : [a, b] \to \mathbf{R}$ be a function of bounded variation. Let $\Lambda(f) = \int_a^b f \, d\lambda$ for every real-valued regulated function f. Show that Λ is a bounded linear functional on the class of regulated functions. That is, $\Lambda(sf + tg) = s\Lambda(f) + t\Lambda(g)$ and $|\Lambda(f)| \leqslant M \|f\|$ for some constant M. ($\|f\| = \text{lub}_x |f(x)|$.)

21. With the same setting as in Exercise 20, think of $\Lambda(f)$ as known for each f. Try to recover λ. (Note that we may let $\lambda(a) = 0$ without impairing generality.) In particular, when $a \leqslant t < b$, find f_t such that $\Lambda(f_t) = \lambda(t+)$. Find f_b such that $\Lambda(f_b) = \lambda(b)$.

22. Let Λ be a bounded linear functional on the regulated functions. Is it of the form $\int_a^b f \, d\lambda$ for a well-chosen λ? Use the results of the preceding exercise to define a candidate λ.

 (a) Show that $\Lambda(f) = \int_a^b f \, d\lambda$ at least for continuous f.

 (b) Let $\Lambda(f) = f(a+) - f(a)$. Show that Λ is a bounded linear functional on the regulated functions. Define λ as in part (a). Does $\Lambda(f) = \int_a^b f \, d\lambda$ for functions f which are not continuous at a?

23. Look up the Hahn-Banach theorem if you are not familiar with it. Use it to prove the Riesz representation theorem:

    Let Λ be a bounded linear functional on the continuous real-valued functions on $[a, b]$. There exists a function of bounded variation λ such that $\Lambda(f) = \int_a^b f \, d\lambda$ for every continuous f.

24. Let α be a function of bounded variation. Suppose $\int_a^b f \, d\alpha = 0$ for every continuous f. Show α is constant.

25. Let α be the primitive of a regulated function on $[a, b]$. Suppose $\int_a^b f$ and $\int_a^b |f|$ exist. Show that $\int_a^b f \, d\alpha$ also exists.

26. Suppose $\int_a^b f \, d\alpha$ exists. Let α be a primitive on $[a, b]$ and $\tau : [c, d] \to [a, b]$ a primitive on $[c, d]$. Suppose further that $\tau'(t) \neq 0$ with countably many exceptions. Set $F(x) = \int_a^x f \, d\alpha$ and assume F is the primitive of $f\alpha'$. Show that $\int_c^d f \circ \tau \, d(\alpha \circ \tau)$ exists and equals $\int_{\tau(c)}^{\tau(d)} f \, d\alpha$.

27. Let α be defined on $[a, b]$. Let β be the restriction of α to $[a, b)$, i.e., β has $[a, b)$ as its domain and $\alpha(x) = \beta(x)$ when $a \leqslant x < b$. Under what conditions is existence of $\int_a^b f \, d\alpha$ equivalent to existence of $\int_a^b f \, d\beta$? What is the relation of the values of the two integrals?

CHAPTER 8

COMPARISON OF INTEGRALS

In preceding chapters the generalized Riemann integral has been presented on its own terms with explicit reference only to the Riemann integral. Nevertheless, the theory of the Lebesgue integral has been in the background at all times. At many points it has given guidance as to what could be proved. At some points the methods of proof were drawn from the Lebesgue theory, too. The existing literature on integration centers on the Lebesgue theory. Thus it is very helpful to know how the generalized Riemann integral ties in with the Lebesgue integral in order to be able to use other books.

Measure and measurable sets as defined in Section 4.6 are identical to Lebesgue measure and Lebesgue measurable sets. The absolutely integrable functions are identical to the Lebesgue integrable functions.

The key to comparing the two senses of measure and measurable sets lies in characterizing measurable sets in a fashion independent of the gauge limit. This is done in Section S8.1 in terms of sets which are the intersection of a sequence of open sets. (They are called G_δ sets.) The crucial element in this discussion is the covering lemma which was given in Section S5.5.

The Lebesgue measurable sets have exactly the same characterization as the one given in Section S8.1.

Knowing that the measurable sets are the same in the two theories, one must go on to see that the measures are identical. This is carried out in Section S8.2. To see that the integrals also agree, one begins with characteristic functions of measurable sets of finite measure, moves on to simple functions, and deals with the general case by applying the monotone convergence theorems in both theories to sequences of simple functions. This is completed in Section S8.2.

Since it turns out that the generalized Riemann integral is applicable to more functions than the Lebesgue integral, one is left to wonder whether exactly the Lebesgue integrable functions can be selected by a definition of Riemann type. E. J. McShane has answered this question affirmatively. (See [8].) His definition uses the gauge limit but applies it to an enlarged collection of sums. His sense of tagged division does not require that the tag belong to the interval of which it is the tag. In Section S8.3 it is shown that absolute integrability is equivalent to integration in McShane's sense. With the prior equivalence of absolute integrability to Lebesgue integrability, it follows that McShane's definition yields integration in the Lebesgue sense, too.

The material of the preceding chapters opens the door to many other topics in integration theory. Section 8.4 lists some of them and suggests some other books which offer further information. These books have been chosen with an eye to their readability on the basis of the background afforded by the present volume.

S8.1. Characterization of measurable sets. The most intricate steps needed to characterize measurable sets in terms of simpler classes of sets have already been taken in Section S5.5. There it was found that each integrable set E

can be covered by a sequence $(J_n)_{n=1}^\infty$ of bounded closed
intervals so that $\sum_{n=1}^\infty M(J_n) < \mu(E) + \epsilon$. Each of these
closed intervals can be enclosed in an open interval H_n
so that $M(H_n) < M(J_n) + \epsilon/2^n$. Let $H = \bigcup_{n=1}^\infty H_n$.
Then $E \subseteq H$ and H is open. Moreover, $\sum_{n=1}^\infty M(H_n)$
$< \sum_{n=1}^\infty M(J_n) + \epsilon$. Consequently, $\mu(E) \leqslant \mu(H) \leqslant$
$\sum_{n=1}^\infty M(H_n) < \mu(E) + 2\epsilon$.

The M-null sets can be characterized now in terms of
sequences of open intervals.

Definition.   A set E is *Lebesgue null* when there exists a
sequence $(H_n)_{n=1}^\infty$ of bounded open intervals such that
$E \subseteq \bigcup_{n=1}^\infty H_n$ and $\sum_{n=1}^\infty M(H_n) < \epsilon$.

A set is M-null if and only if it is Lebesgue null.

When E is M-null $\mu(E) = 0$. Thus existence of the
covering sequence with small total length is assured by the
conclusion at the end of the first paragraph of this section.

Now suppose E is Lebesgue null. To show that E is
M-null take a covering sequence $(G_n)_{n=1}^\infty$ and let $\gamma(z)$ be
the first G_n containing z when $z \in E$. Let $\mathcal{D}$ be a γ-fine
partial division whose tags are in E. Let $\mathcal{D}_n$ be the subset
of $\mathcal{D}$ consisting of all zJ such that $\gamma(z) = G_n$. The union of
the intervals in $\mathcal{D}_n$ is a subset of G_n. Thus $M(\mathcal{D}_n)$
$\leqslant M(G_n)$ and $M(\mathcal{D}) \leqslant \sum_{n=1}^\infty M(G_n) < \epsilon$. Therefore E is
M-null.

There is one more preliminary step before the
characterization of measurable sets can be done.

*If E is measurable there is an open set H such that
$E \subseteq H$ and $\mu(H - E) < \epsilon$.*

Take expanding bounded intervals I_n so that $\mathbf{R}^p =$
$\bigcup_{n=1}^\infty I_n$. Let $E_n = E \cap I_n$. Then E_n is integrable and there
is an open set H_n such that $E_n \subseteq H_n$ and $\mu(H_n) < \mu(E_n) +$

$\epsilon/2^n$. Now $\mu(H_n - E_n) = \mu(H_n) - \mu(E_n) < \epsilon/2^n$. Set $H = \bigcup_{n=1}^{\infty} H_n$. Any union of open sets is open. Thus H is open, $E \subseteq H$, and $H - E \subseteq \bigcup_{n=1}^{\infty}(H_n - E_n)$. Consequently $\mu(H - E) \leqslant \sum_{n=1}^{\infty} \mu(H_n - E_n) < \epsilon$.

One more class of sets must be introduced. A set is a G_δ set provided it is the intersection of a sequence of open sets. Of course every open set is a G_δ set. Not all G_δ sets are open. Example: any finite set is a G_δ set but not open.

A set E is measurable if and only if there is a G_δ set B such that $E \subseteq B$ and $B - E$ is Lebesgue null.

Suppose E is measurable. Using the preliminary result, cover E by open sets H_n such that $\mu(H_n - E) < 1/n$ for all n. Set $B = \bigcap_{n=1}^{\infty} H_n$. Then $E \subseteq B \subseteq H_n$ and $B - E \subseteq H_n - E$. Thus $\mu(B - E) < 1/n$ for all n. Consequently, $\mu(B - E) = 0$. This makes $B - E$ Lebesgue null as well.

Conversely, a Lebesgue null set $B - E$ is M-null and thus measurable. The measurable sets are closed under the formation of countable intersections. Thus every G_δ set B is measurable. Since $E = B - (B - E)$, E is measurable.

Now we have a criterion which allows us to distinguish between measurable and nonmeasurable sets without recourse to the gauge limit.

S8.2. Lebesgue measure and integral. There are many ways to define and develop the Lebesgue integral. Some of them, including Lebesgue's original presentation, begin with measure and proceed to the integral. Others reverse the process. They all end with the same class of measurable sets, the same measure, and the same integral. For the purpose of comparing the Lebesgue integral with the generalized Riemann integral it is not necessary to go through any version in detail. It will suffice to list and use a few basic facts about Lebesgue's measure and integral.

What are the basic facts of Lebesgue measure? The definition of Lebesgue null set was given above. These are the sets to which Lebesgue measure assigns the value zero. Secondly, a set E is Lebesgue measurable if and only if there is a G_δ set B such that $E \subseteq B$ and $B - E$ is a Lebesgue null set. Thus the measurable sets derived from the generalized Riemann integral are identical with the Lebesgue measurable sets. Let m denote the Lebesgue measure function. We know so far that m and μ have the same domain. Does $m(E) = \mu(E)$ for every measurable set E? Lebesgue measure assigns to a bounded interval I the value $M(I)$. Moreover, m is countably additive. When H is open there is a sequence of nonoverlapping bounded intervals $(I_n)_{n=1}^\infty$ such that $H = \bigcup_{n=1}^\infty I_n$. Thus $m(H) = \sum_{n=1}^\infty M(I_n)$. As we know, this sum is also $\mu(H)$. In general, when E is any measurable set, $m(E)$ is the greatest lower bound of $m(H)$ for all open H containing E. Since $\mu(E) \leqslant \mu(H)$ and $\mu(H) = m(H)$, $\mu(E) \leqslant m(E)$ for every E. Thus $m(E) = \infty$ when $\mu(E) = \infty$. When $\mu(E) < \infty$, there are open sets H_n such that $E \subseteq H_n$ and $\mu(H_n) < \mu(E) + 1/n$. Thus $m(E) \leqslant \mu(E)$. Again $\mu(E) = m(E)$.

Now consider the Lebesgue integral $(\mathcal{L})\int_I f$. The basic fact is that $m(E)$ is the integral of the characteristic function of a set E of finite measure. Thus the two integrals agree in this case. The Lebesgue integral is linear, too. Thus the two integrals agree on simple functions which take nonzero values on sets of finite measure.

Measurable functions in the Lebesgue theory agree with the ones defined in Section 5.1 since the two classes of measurable sets are identical. The Lebesgue integrable functions are a subclass of the measurable functions, as the absolutely integrable functions are. We shall show that they are the same and that the integrals agree on this class.

Let f be a nonnegative measurable function. There is a sequence of simple functions f_n increasing to f. Each f_n vanishes outside a bounded interval. (See p. 139.) Thus $\int_I f_n = (\mathcal{L})\int_I f_n$. Both integration theories have a monotone convergence theorem in which boundedness of the sequence of integrals is equivalent to integrability of the limit function. Existence of $\int_I f$ or $(\mathcal{L})\int_I f$ implies boundedness of the sequence of integrals. Thus existence of $\int_I f$ is equivalent to existence of $(\mathcal{L})\int_I f$. Moreover, $\int_I f = \lim_{n\to\infty}\int_I f_n$ and likewise for Lebesgue integrals. Thus $\int_I f = (\mathcal{L})\int_I f$.

A measurable function f of variable sign is Lebesgue integrable if and only if f^+ and f^- are both Lebesgue integrable. Likewise generalized Riemann integrability of f and $|f|$ is equivalent to integrability of f^+ and f^-. The identity of the integrals on nonnegative functions thus permits the desired conclusion for real-valued functions.

Finally, vector-valued functions can be treated by consideration of real components.

S8.3. Characterization of absolute integrability using Riemann sums.

First, let us make a general observation about limits. As an illustration, recall that the right-hand limit $\lim_{x\to a^+} f(x)$ may exist when $\lim_{x\to a} f(x)$ does not exist. The reason is that the right-hand limit ignores some function values which are required to cluster around the limiting value in the two-sided limit. The same principle accounts for the difference in generality between the Riemann integral and the generalized Riemann integral. In the definition of the former, all tagged divisions with lengths of subintervals less than some fixed δ are required to give Riemann sums which cluster around the limiting value. The generalized Riemann definition tests the tagged division against a gauge γ. Since γ need have no smallest interval, many Riemann sums which are required to compete in the Riemann definition fail to qualify for consideration in the generalized Riemann definition.

The same principle can be used to identify the absolutely integrable functions within the totality of the functions integrable in the generalized Riemann sense. Since the absolutely integrable functions are a subclass of the integrable functions, we seek a limit process which requires more function values, i.e., Riemann sums, to cluster about the limit than the generalized Riemann integral does. E. J. McShane's idea is to broaden the concept of tagged division while using the gauge limit as before. Note how this differs from the distinction between the Riemann and the generalized Riemann integral. Only one notion of tagged division is used in the two integrals but different limits are applied.

In the broader sense introduced by McShane, a tagged subinterval of I consists of a point z in I and a closed interval J with $J \subseteq I$. The tag z is no longer required to be in J. As before, zJ is γ-fine provided $J \subseteq \gamma(z)$. For the sake of distinguishing the two senses of tagged division, a tagged division in the broader sense will be referred to as an $\mathfrak{M}$-division. The same prefix will indicate concepts associated with $\mathfrak{M}$-divisions. When no distinguishing letter is used, the original sense is to be understood.

Since every γ-fine division is also a γ-fine $\mathfrak{M}$-division, a function which is $\mathfrak{M}$-integrable is also integrable and the two integrals are equal. There are two major goals to be achieved. It must be shown that $|f|$ is $\mathfrak{M}$-integrable whenever f is $\mathfrak{M}$-integrable. It must also be shown that integrability of f and $|f|$ implies $\mathfrak{M}$-integrability of f.

Together these characterize absolute integrability as $\mathfrak{M}$-integrability.

It is not necessary to develop the $\mathfrak{M}$-integral extensively. We will need the Cauchy criterion, linearity, and Henstock's lemma. The proofs of these properties of the $\mathfrak{M}$-integral are the same as the ones for the generalized Riemann integral.

From the Cauchy criterion we see that f must vanish at all infinite points in order to be $\mathfrak{M}$-integrable.

The argument for $\mathfrak{M}$-integrability of $|f|$ when f is $\mathfrak{M}$-integrable which follows is the one given by McShane in [8]. The next proposition is the key.

Let $|\int_I f - fM(\mathfrak{D})| < \epsilon$ for every γ-fine $\mathfrak{M}$-division $\mathfrak{D}$. Let $\mathfrak{D}$ and $\mathcal{E}$ be such divisions. Then

$$\sum |f(y) - f(z)| M(J \cap K) < 2q\epsilon$$

where the summation is over all yJ in $\mathfrak{D}$ and zK in $\mathcal{E}$.

First let $q = 1$. The collection $\mathfrak{F}$ defined by

$$\mathfrak{F} = \{ J \cap K : yJ \in \mathfrak{D} \quad \text{and} \quad zK \in \mathcal{E} \}$$

is made up of nonoverlapping closed intervals. The union of the nondegenerate intervals in $\mathfrak{F}$ is I. Form two $\mathfrak{M}$-divisions $\mathfrak{D}'$ and $\mathcal{E}'$ from $\mathfrak{F}$ and the tags of $\mathfrak{D}$ and $\mathcal{E}$ as follows. The tag y' of $J \cap K$ in $\mathfrak{D}'$ is y when $f(y) \geqslant f(z)$ and z when $f(z) > f(y)$. The remaining one of y and z is the tag z' of $J \cap K$ in $\mathcal{E}'$. Thus

$$f(y') - f(z') = |f(y) - f(z)|$$

and

$$fM(\mathfrak{D}') - fM(\mathcal{E}') = \sum |f(y) - f(z)| M(J \cap K).$$

Both $\mathfrak{D}'$ and $\mathcal{E}'$ are γ-fine since both $\mathfrak{D}$ and $\mathcal{E}$ are

γ-fine. Consequently, $|fM(\mathcal{D}') - fM(\mathcal{E}')| < 2\epsilon$. This completes the argument when $q = 1$.

The length of a vector never exceeds the sum of the absolute values of its components. From this fact the q-dimensional case follows at once.

The proposition just completed and the Cauchy criterion yield $\mathfrak{M}$-integrability of $|f|$.

Start with γ as above and let $\mathcal{D}$ and $\mathcal{E}$ be γ-fine. Form $\mathcal{D}''$ and $\mathcal{E}''$ from $\mathcal{F}$ and the tags of $\mathcal{D}$ and $\mathcal{E}$ by putting $y(J \cap K)$ in $\mathcal{D}''$ and $z(J \cap K)$ in $\mathcal{E}''$ when yJ is in $\mathcal{D}$ and zK is in $\mathcal{E}$. Then $|f|M(\mathcal{D}) = |f|M(\mathcal{D}'')$ and $|f|M(\mathcal{E}) = |f|M(\mathcal{E}'')$. (Note that the fact that f vanishes on infinite points enters here.) But

$$|f(y)|M(J \cap K) - |f(z)|M(J \cap K)$$
$$\leqslant |f(y) - f(z)|M(J \cap K).$$

Thus $|f|M(\mathcal{D}'') - |f|M(\mathcal{E}'') < 2q\epsilon$. The same reasoning applies with $\mathcal{D}''$ and $\mathcal{E}''$ interchanged. Thus $||f|M(\mathcal{D}'') - |f|M(\mathcal{E}'')| < 2q\epsilon$. Since the primes can be dropped, the Cauchy criterion is satisfied for $\mathfrak{M}$-integrability of $|f|$.

The argument from absolute integrability to $\mathfrak{M}$-integrability goes in steps. Start with χ_E where E is a bounded measurable set. Fix a closed bounded interval J such that $E \subseteq J$. Let $F = J - E$. There are open sets G and H containing E and F, respectively, such that $\mu(G) < \mu(E) + \epsilon$ and $\mu(H) < \mu(F) + \epsilon$. Define a gauge γ so that $\gamma(z) \subseteq G$ when $z \in E$, $\gamma(z) \subseteq H$ when $z \in F$, and $J \cap \gamma(z)$ is empty when $z \notin J$. Let $\mathcal{D}$ be a γ-fine $\mathfrak{M}$-division. Let $\mathcal{D}_E$ be the subset of $\mathcal{D}$ whose tags are in E. The intervals of $\mathcal{D}_E$ are nonoverlapping and they are all contained in G. Thus

$$\chi_E M(\mathcal{D}) = M(\mathcal{D}_E) \leqslant \mu(G) < \mu(E) + \epsilon.$$

Similarly $\chi_F M(\mathcal{D}) < \mu(F) + \epsilon$. Since $J \cap \gamma(z)$ is empty when $z \notin J$, the intervals of $\mathcal{D}$ with tags in J cover J. Hence $\chi_J M(\mathcal{D}) \geqslant \mu(J)$. Since $\chi_J = \chi_E + \chi_F$,

$$\chi_E M(\mathcal{D}) = \chi_J M(\mathcal{D}) - \chi_F M(\mathcal{D})$$
$$> \mu(J) - (\mu(F) + \epsilon)$$
$$\geqslant \mu(E) - \epsilon.$$

Now $|\mu(E) - \chi_E M(\mathcal{D})| < \epsilon$. Hence χ_E is $\mathcal{M}$-integrable.

The linearity of the $\mathcal{M}$-integral insures that every simple function which takes its nonzero values on bounded sets is $\mathcal{M}$-integrable.

The final step up to any absolutely integrable function can be done either by a direct calculation or by use of convergence theorems for the $\mathcal{M}$-integral (which must be verified first). The direct calculation follows.

Let f be absolutely integrable on $\overline{\mathbf{R}}^p$ with zero values at the infinite points. Let $I = \overline{\mathbf{R}}^p$, $I_n = [-n, n] \times [-n, n] \times \cdots \times [-n, n]$, and set $E_n = I_n \cap f^{-1}([-n, n])$. Construct simple functions f_n which vanish outside I_n and satisfy $|f(x) - f_n(x)| < 1/2^n$ for all x in E_n. Note that $E_n \subseteq E_{n+1}$ and $\bigcup_{n=1}^{\infty} E_n = \mathbf{R}^p$. We can also require that $|f_n| \leqslant |f|$. Thus $|f - f_n| \leqslant 2|f|$ and $\lim_{n \to \infty} |f - f_n| = 0$ everywhere. The dominated convergence theorem of the generalized Riemann integral allows us to choose an integer N such that $n < N$ and $\int_I |f - f_N| < 1/2^n$. Fix such an N for each n. (Strictly speaking we should write N_n.)

Let γ_n be a gauge on I such that $|\int_I f_N - f_N M(\mathcal{D})| < 1/2^n$ when $\mathcal{D}$ is any γ_n-fine $\mathcal{M}$-division. (Remember that each of these simple functions is already known to be $\mathcal{M}$-integrable.) Since I_n is contained in the interior of I_{n+1}, we may also require that $\gamma_n(z)$ be contained in I_{n+1} when $z \in I_n$.

Our objective is to define a gauge γ on I so that $|\int_I f - f M(\mathcal{D})| < \epsilon$ when $\mathcal{D}$ is any γ-fine $\mathcal{M}$-division. (The assumed existence of $\int_I f$ allows the choice of γ for ordinary tagged divisions $\mathcal{D}$. Extra restrictions on γ will be needed to allow inclusion of all $\mathfrak{M}$-divisions.) We construct γ out of the gauges γ_n. We will use only γ_n for $n \geqslant m$ and specify later the appropriate value of m.

Let $\gamma(z) = \gamma_m(z)$ on E_m. For $n > m$ let $\gamma(z) = \gamma_n(z)$ on $E_n - E_{n-1}$. This defines γ only on $\mathbf{R}^p$. Choose an integer Q so that $\int_I |f| - \int_{I_Q} |f| < \epsilon/2$. On $\overline{\mathbf{R}}^p - \mathbf{R}^p$ define $\gamma(z)$ so that $\gamma(z)$ does not intersect I_Q.

Now let $\mathcal{D}$ be a γ-fine $\mathfrak{M}$-division. Set $\mathcal{D}_n = \{ zJ \in \mathcal{D} : \gamma(z) = \gamma_n(z) \}$ for $n \geqslant m$ and let $\mathcal{D}_\infty$ be the subset of $\mathcal{D}$ whose tags are infinite points. Then $fM(\mathcal{D}_\infty) = 0$ since f vanishes on infinite points. For convenience let $\nu(J)$, $\nu_n(J)$, $\rho(J)$ and $\rho_n(J)$ be the integrals over J of f, f_n, $|f|$, and $|f - f_n|$, respectively. Now

$$\int_I f - fM(\mathcal{D}) = \nu(\mathcal{D}_\infty) + \sum_{n=m}^{\infty} \left[\nu(\mathcal{D}_n) - fM(\mathcal{D}_n) \right].$$

The intervals of $\mathcal{D}_\infty$ are contained in $I - I_Q$. Moreover $|\nu(J)| \leqslant \rho(J)$ for all J. Thus

$$|\nu(\mathcal{D}_\infty)| \leqslant \rho(\mathcal{D}_\infty) \leqslant \rho(I - I_Q) = \rho(I) - \rho(I_Q) < \epsilon/2.$$

To estimate the summands we do some obvious addition and subtraction to get

$$
\begin{aligned}
|\nu(\mathcal{D}_n) - fM(\mathcal{D}_n)| \leqslant\ & |\nu(\mathcal{D}_n) - \nu_N(\mathcal{D}_n)| \\
& + |\nu_N(\mathcal{D}_n) - f_N M(\mathcal{D}_n)| \\
& + |f_N M(\mathcal{D}_n) - fM(\mathcal{D}_n)|.
\end{aligned}
$$

For the first term we have

$$|\nu(\mathcal{D}_n) - \nu_N(\mathcal{D}_n)| \leqslant \rho_N(\mathcal{D}_n) \leqslant \int_I |f - f_N| < 1/2^n$$

because of the choice of N. Henstock's lemma for the $\mathfrak{M}$-integral gives $|\nu_N(\mathcal{D}_n) - f_N M(\mathcal{D}_n)| \leqslant 1/2^n$ because $\mathcal{D}_n$

is a γ_n-fine subset of an $\mathfrak{M}$-division. On the third term we use the fact that f_N approximates f with known error on E_N. Since $n \leqslant N$, $E_n \subseteq E_N$. Every tag of $\mathfrak{D}_n$ is in E_n. Moreover, every interval of $\mathfrak{D}_n$ is contained in I_{n+1}. Thus

$$|f_N M(\mathfrak{D}_n) - fM(\mathfrak{D}_n)| \leqslant |f_N - f| M(\mathfrak{D}_n) \leqslant 2^{-N} M(I_{n+1}).$$

But $2^{-N} \leqslant 2^{-n}$ and $M(I_{n+1}) = 2^p (n+1)^p$. The combined result for the summand is

$$|\nu(\mathfrak{D}_n) - fM(\mathfrak{D}_n)| < 1/2^n + 1/2^n + 2^p (n+1)^p / 2^n.$$

The final estimate is

$$\left| \int_I f - fM(\mathfrak{D}) \right| < \epsilon/2 + \sum_{n=m}^{\infty} (2 + 2^p (n+1)^p)/2^n.$$

Now choose m so that the last series has a value less than $\epsilon/2$. This completes the argument.

This argument throws very little light on how $\mathfrak{M}$-divisions rule out integrability based on the delicate interplay of positive and negative values. The crucial difference seems to be that there is no limitation on the number of intervals which can have a common tag in $\mathfrak{M}$-divisions. When the tag must belong to its interval the maximum number of intervals with a common tag is 2^p in $\overline{\mathbf{R}}^p$. The following example helps to show how $\mathfrak{M}$-integrability fails.

Example 1. Consider again the function f defined on $[0, 1]$ as follows. Let $0 = c_0 < c_1 < c_2 < \cdots$ with $\lim_{n \to \infty} c_n = 1$. Let $(a_n)_{n=1}^{\infty}$ be the sequence $1, -1/2, 1/2, -1/3, 1/3, \ldots$. When $c_{n-1} \leqslant x < c_n$ set $f(x) = a_n/(c_n - c_{n-1})$. Set $f(1) = 0$. Show directly from the definition that f is not $\mathfrak{M}$-integrable.

Solution. Recall that f has a primitive F such that $F(0) = 0$, $F(1) = 1$, and $\Delta F(B_n) = a_n$ when $B_n = [c_{n-1}, c_n]$.

Thus $\int_0^1 f = 1$. It suffices to show that for every gauge γ there is a γ-fine $\mathfrak{M}$-division $\mathcal{D}$ such that $fL(\mathcal{D}) > 2$. This will be achieved by tagging a large number of intervals of $\mathcal{D}$ with 1.

Let γ be given. For each n fix a γ-fine division $\mathcal{E}_n$ of B_n such that $|\Delta F(B_n) - fL(\mathcal{E}_n)| < 1/2^n$. Choose an odd integer M so that $c_M \in \gamma(1)$. Let N be an even integer larger than M. Let V be the set of odd integers between M and N, i.e., $V = \{M + 2, M + 4, \ldots, N - 1\}$. The numbers a_n for odd n are the terms of the harmonic series. Thus we can impose on N the further requirement that $\sum_{n \in V} a_n > 2$. Let $\mathcal{D}_1 = \bigcup_{n=1}^{M} \mathcal{E}_n$ and $\mathcal{D}_2 = \bigcup_{n \in V} \mathcal{E}_n$. The subintervals of $[0, 1]$ not covered by $\mathcal{D}_1$ and $\mathcal{D}_2$ are the even-numbered intervals B_n with $n = M + 1, M + 3, \ldots, N$ and the interval $[c_N, 1]$. Tag each of these with 1 and call the resulting collection $\mathcal{D}_3$. Set $\mathcal{D} = \mathcal{D}_1 \cup \mathcal{D}_2 \cup \mathcal{D}_3$. Then $\mathcal{D}$ is clearly a γ-fine $\mathfrak{M}$-division of $[0, 1]$. We propose to show that $fL(\mathcal{D}) > 2$.

When n is even $a_n + a_{n+1} = 0$. Thus $\Delta F(\mathcal{D}_1) = \sum_{n=1}^{M} a_n = a_1 = 1$. Moreover $\Delta F(\mathcal{D}_2) = \sum_{n \in V} a_n > 2$. The choice of $\mathcal{E}_n$ implies that

$$|fL(\mathcal{D}_1 \cup \mathcal{D}_2) - \Delta F(\mathcal{D}_1 \cup \mathcal{D}_2)| < \sum_{n=1}^{N} 1/2^n < 1.$$

Since $f(1) = 0$ and every interval in $\mathcal{D}_3$ is tagged with 1, $fL(\mathcal{D}_3) = 0$. Now

$$\begin{aligned} fL(\mathcal{D}) &= fL(\mathcal{D}_1 \cup \mathcal{D}_2) \\ &= \Delta F(\mathcal{D}_1) + \Delta F(\mathcal{D}_2) + (fL(\mathcal{D}_1 \cup \mathcal{D}_2) \\ &\quad - \Delta F(\mathcal{D}_1 \cup \mathcal{D}_2)) \\ &> 1 + 2 - 1, \end{aligned}$$

as promised.

At first glance it might appear crucial to have $f(1) = 0$

in Example 1. This is not so. The only change one need make is to require that $\sum_{n \in V} a_n > 2 + |f(1)|$ in order to have a proof valid for any choice of $f(1)$.

8.4 Suggestions for further study. Since the Lebesgue integral is identical with the generalized Riemann integral on the absolutely integrable functions, this treatment of integration opens the door to the many treatises on Lebesgue integration. There are two basically different approaches. One begins with measure and proceeds to the integral. Williamson [12] is a concise, lucid account of it. Munroe [10] is a more general treatment. The reverse procedure is followed by McShane and Botts [9]. Taylor [11] gives an introduction to each of these approaches. There are many variants of each approach; thus the definition and development of the Lebesgue integral may differ considerably from book to book even within a single approach.

There are large portions of integration theory which have been left untouched in the foregoing pages. A few topics of particular utility will be mentioned in the remaining paragraphs.

Differentiation theory of functions and integrals is closely tied to integrals. One of the basic facts is the existence of a derivative almost everywhere for a monotone function. See Boas [1] for this. Lebesgue's original presentation of his integral culminated in his proof that the derivative of $\int_a^x f$ is f almost everywhere. Williamson [12] offers a clear and simple discussion. There is also differentiation theory for integrals on higher dimensional spaces $\mathbf{R}^p$. Munroe [10] gives an account of it.

The change of variables in multiple integrals is an important tool. At present it does not appear that the generalized Riemann integral offers any help in reducing the complexities of the proof of this theorem. Williamson [12] offers one of the standard treatments.

The set of measurable functions f such that $|f|^p$ is integrable is called L_p space. Hölder's inequality in Exercise 4 of Section 5.4 is important in L_p space. A brief discussion is given in Williamson [12] and a more extensive one in Munroe [10].

There is much more measure theory. See Munroe [10].

The theory of Fourier series and integrals is one of the parts of analysis most affected by the development of Lebesgue's powerful theory of integration. See "Harmonic Analysis" by Guido Weiss in [5] for a survey and further references.

These topics are only a few of those one might mention. The works cited will point the way to other topics and other books. Munroe [10] offers helpful lists of references by subject at the end of each chapter.

REFERENCES

1. R. P. Boas, Jr., *A Primer of Real Functions*, Carus Mathematical Monographs, no. 13, Mathematical Association of America, 1960.

2. Ralph Henstock, *Linear Analysis*, Butterworths, London, 1967.

3. ———, A Riemann-type integral of Lebesgue power, *Canad. J. Math.*, 20 (1968), 79–87.

4. ———, *Theory of Integration*, Butterworths, London, 1963.

5. I. I. Hirschman, Jr. (ed.), *Studies in Real and Complex Analysis*, Studies in Mathematics, vol. 3, Mathematical Association of America, 1965.

6. Jaroslav Kurzweil, Generalized ordinary differential equations and continuous dependence on a parameter, *Czechoslovak Math. J.*, 7 (82) (1957) 418–446.

7. E. J. McShane, A Riemann-type integral that includes Lebesgue-Stieltjes, Bochner and stochastic integrals, *Mem. Amer. Math. Soc.*, 88 (1969).

8. ———, A unified theory of integration, *Amer. Math. Monthly*, 80 (1973) 349–359.

9. E. J. McShane and Truman Botts, *Real Analysis*, Van Nostrand, Princeton, 1959.

10. M. E. Munroe, *Measure and Integration*, 2nd ed., Addison-Wesley, Reading, Mass., 1971.

11. A. E. Taylor, *General Theory of Functions and Integration*, Blaisdell, Waltham, Mass., 1965.

12. J. H. Williamson, *Lebesgue Integration*, Holt, Rinehart and Winston, New York, 1962.

APPENDIX

SOLUTIONS OF IN-TEXT EXERCISES

Chapter 1

1. Begin by choosing an integer m so that $1/m \leqslant \epsilon$. Enclose each number $1/j$ in an interval $[1/j - d_j, 1/j + d_j]$ which is contained in $\gamma(1/j)$ for $j = 1, 2, \ldots, m$. These can also be chosen so that $1/j + d_j < 1/(j-1) - d_{j-1}$ for $j = 2, 3, \ldots, m$ and $0 < 1/m - d_m$.

Tag $[1/j - d_j, 1/j + d_j]$ with $1/j$ for $2 \leqslant j \leqslant m$. Tag $[1 - d_1, 1]$ with 1. The rest of $[0, 1]$ is $[0, 1/m - d_m]$ and each of the intervals $[1/j + d_j, 1/(j-1) - d_{j-1}]$ for $2 \leqslant j \leqslant m$. Tag the former with 0 and the latter with any of its points, say $1/j + d_j$. Now we have a division of $[0, 1]$ with $2m$ intervals tagged in such a way as to be γ-fine.

2. (a) Let $\mathcal{D} \in R_\delta$ and let $zJ \in \mathcal{D}$. Since J has length less than δ it is contained in any open interval of length 2δ centered on a point of J. Thus $J \subseteq (z - \delta, z + \delta)$. In consequence $\mathcal{D} \in GR_\delta$.

When $\mathcal{D} \in GR_\delta$ and $zJ \in \mathcal{D}$, $J \subseteq (z - \delta, z + \delta)$. Since the length of $(z - \delta, z + \delta)$ is 2δ and the endpoints of J are between $z - \delta$ and $z + \delta$, $L(J) < 2\delta$. Consequently $\mathcal{D} \in R_{2\delta}$.

(b) Suppose the Riemann integral of f on $[a, b]$ exists. Choose δ so that $|\int_a^b f - fL(\mathcal{D})| < \epsilon$ when $L(J) < \delta$ for all

247

zJ in $\mathcal{D}$. That is, the inequality holds for all $\mathcal{D}$ in R_δ. Let $\gamma(z) = (z - \delta/2, z + \delta/2)$ for all z in $[a, b]$. Then the set of all γ-fine $\mathcal{D}$ is $GR_{\delta/2}$. Since $GR_{\delta/2} \subseteq R_\delta$, the generalized Riemann integral of f exists and is the same as its Riemann integral.

Now let f be a generalized Riemann integrable function with the added property that $|\int_a^b f - fL(\mathcal{D})| < \epsilon$ for all $\mathcal{D}$ in GR_δ for some positive δ. Since $R_\delta \subseteq GR_\delta$ the Riemann integral of f also exists.

Now we have characterized the Riemann integrable functions among the generalized Riemann integrable functions as those for which there is a gauge $\gamma(z) = (z - \delta, z + \delta)$ with constant δ such that $|\int_a^b f - fL(\mathcal{D})| < \epsilon$ for all γ-fine $\mathcal{D}$.

3. Let $f(x) = 0$ when x is irrational. Let p/q be a fraction in lowest terms. Set $f(p/q) = q$. Since the rationals are countable this function is integrable on any interval $[a, b]$ and its integral is zero, according to Example 4, p. 19. Also f is unbounded on every interval $[c, d]$ for the following reasons. Let M be given. Select a prime number q such that $q > M$ and $2/q < d - c$. There is an integer k such that $k/q \leqslant c < (k + 1)/q$. Then $(k + 2)/q < d$. The prime q divides at most one of $k + 1$ and $k + 2$. Thus $f((k + 1)/q) = q$ or $f((k + 2)/q) = q$.

4. There are elements a and b in $\overline{\mathbf{R}}$ such that I is one of (a, b), $[a, b)$, $(a, b]$, and $[a, b]$. Fix d in (a, b). There is a function f on I such that F and G are primitives of f on I, hence on any *closed* subinterval of I. For any x in I such that $x < d$, $\int_x^d f = F(d) - F(x)$ and $\int_x^d f = G(d) - G(x)$. Thus $F(x) = G(x) + K$ where $K = F(d) - G(d)$ when $x \in I$ with $x < d$. When $x \in I$ and $d < x$, $\int_d^x f = F(x) - F(d) = G(x) - G(d)$. Again $F(x) = G(x) + K$. Trivially $F(d) = G(d) + K$. Thus $F(x) = G(x) + K$ for all x in I.

5. There is a gauge γ on $[a, b]$ such that $|\int_a^b f - fL(\mathcal{D})| < \epsilon$ when $\mathcal{D}$ is a γ-fine division of $[a, b]$. Let $s \in (a, b)$.

Since f is also integrable on $[s, b]$, there is a gauge γ_s such that $|\int_s^b f - fL(\mathcal{E})| < \epsilon$ when $\mathcal{E}$ is a γ_s-fine division of $[s, b]$. It is possible to choose γ_s so that $\gamma_s(z) \subseteq \gamma(z)$, too. Choose c so that $c \in \gamma(a)$ and $|f(a)|L([a, c]) < \epsilon$. Let $s \in (a, c)$. Let $\mathcal{E}$ be a γ_s-fine division of $[s, b]$. Let $\mathcal{D} = \{a[a, s]\} \cup \mathcal{E}$. Then $\mathcal{D}$ is a γ-fine division of $[a, b]$. Now

$$\left| \int_a^b f - \int_s^b f \right| \leqslant \left| \int_a^b f - fL(\mathcal{D}) \right| + \left| fL(\mathcal{E}) - \int_s^b f \right|$$

$$+ |f(a)L([a, s])|.$$

Each term on the right is less than ϵ. Hence $\lim_{s \to a} \int_s^b f = \int_a^b f$.

6. We know that $\int_0^\infty f$ exists if and only if $\lim_{t \to \infty} \int_0^t f$ exists. Also $\int_0^\infty f = \lim_{t \to \infty} \int_0^t f$. Using a primitive of f on $[0, t]$ we get $\int_0^t f = \sum_{k=1}^m a_k + (t - m)a_{m+1}$ where $m \leqslant t < m + 1$. When $\int_0^\infty f$ exists we specialize t to integer values and get $\int_0^\infty f = \lim_{m \to \infty} \int_0^m f = \sum_{k=1}^\infty a_k$. Conversely, when the series converges, $\lim_{m \to \infty} a_{m+1} = 0$. Thus $\lim_{t \to \infty} \int_0^t f = \lim_{m \to \infty} \sum_{k=1}^m a_k = \sum_{k=1}^\infty a_k$.

7. (a) This is the scale:

$$\xleftarrow{\;\;\;\;} \quad -\infty \;\; -9 \; -4 \; -2 \;\; -1 \; -1/2 \qquad 0 \qquad 1/2 \;\; 1 \;\; 2 \;\; 4 \;\; 9 \;\; \infty$$

(b) Clearly h is strictly increasing and maps $[-\infty, \infty]$ onto $[-1, 1]$. Given x and y in $\bar{\mathbf{R}}^p$ with $x \neq y$, there is some coordinate where they differ, say $x_i \neq y_i$. Then $h(x_i) \neq h(y_i)$ so that $H(x) \neq H(y)$. We have shown H is one-to-one. To show that it is onto, take y such that $-1 \leqslant y_i \leqslant 1$ for $1 \leqslant i \leqslant p$. There is x_i in $\bar{\mathbf{R}}$ such that $h(x_i) = y_i$. Hence H maps the point x with these components x_i onto y.

Since h is increasing it maps $[u, v]$ onto $[h(u), h(v)]$. Consequently H maps $[u_1, v_1] \times \cdots \times [u_p, v_p]$ onto the Cartesian product of the intervals $[h(u_i), h(v_i)]$. The same is true of intervals other than closed intervals.

(c) Draw a square. Put on its edges the scale shown in part (a).

Chapter 2

1. Fix γ_1 and γ_2 so that $|fM(\mathcal{D}) - \int_I f| < \epsilon$ when $\mathcal{D}$ is γ_1-fine and $|gM(\mathcal{D}) - \int_I g| < \epsilon$ when $\mathcal{D}$ is γ_2-fine. Let $\gamma(z) \subseteq \gamma_1(z) \cap \gamma_2(z)$ for all z in $[a, b]$. Then $\mathcal{D}$ is γ_1-fine and γ_2-fine whenever it is γ-fine. Thus

$$\left| (f + g)M(\mathcal{D}) - \int_I f - \int_I g \right|$$

$$\leqslant \left| fM(\mathcal{D}) - \int_I f \right| + \left| gM(\mathcal{D}) - \int_I g \right| < 2\epsilon$$

when $\mathcal{D}$ is γ-fine since $(f + g)M(\mathcal{D}) = fM(\mathcal{D}) + gM(\mathcal{D})$. This shows that $f + g$ is integrable and that $\int_I f + \int_I g$ is its integral.

Since $(cf)M(\mathcal{D}) = c(fM(\mathcal{D}))$ we also have

$$\left| (cf)M(\mathcal{D}) - c\int_I f \right| \leqslant |c| \left| fM(\mathcal{D}) - \int_I f \right| \leqslant |c|\epsilon$$

when $\mathcal{D}$ is γ-fine. Thus $c\int_I f$ is the integral of cf.

A standard induction argument shows that $\sum_{k=1}^n c_k f_k$ is integrable when each f_k is integrable and that $\sum_{k=1}^n c_k \int_I f_k$ is its integral.

2. Let γ be a gauge on I such that $|\int_I f - fM(\mathcal{D})| < \epsilon$ and $|\int_I g - gM(\mathcal{D})| < \epsilon$ when $\mathcal{D}$ is γ-fine. Since g is real valued, $gM(\mathcal{D}) < \int_I g + \epsilon$. Also $|fM(\mathcal{D})| \leqslant gM(\mathcal{D})$ for all

$\mathcal{D}$ since $|f| \leqslant g$ and M takes nonnegative values. Thus

$$\left| \int_I f \right| \leqslant |fM(\mathcal{D})| + \left| \int_I f - fM(\mathcal{D}) \right|$$

$$\leqslant gM(\mathcal{D}) + \epsilon < \int_I g + 2\epsilon.$$

Since ϵ is arbitrary, $|\int_I f| \leqslant \int_I g$, as claimed.

3. By hypothesis, for each whole number n there is a gauge γ_n such that $|fM(\mathcal{D}) - fM(\mathcal{E})| < 1/n$ when $\mathcal{D}$ and $\mathcal{E}$ are γ_n-fine divisions of I. We may also suppose $\gamma_j(z) \subseteq \gamma_i(z)$ when $i < j$. (Replace $\gamma_n(z)$ by $\gamma_1(z) \cap \gamma_2(z) \cap \cdots \cap \gamma_n(z)$.) For each n fix a γ_n-fine division $\mathcal{D}_n$.

Consider the sequence of elements $fM(\mathcal{D}_n)$ in $\mathbf{R}^q$. Let's show that this is a Cauchy sequence. Suppose $i < j$. Then $\mathcal{D}_j$ is not only γ_j-fine it is also γ_i-fine since γ_j is stricter than γ_i. Hence

$$|fM(\mathcal{D}_j) - fM(\mathcal{D}_i)| < 1/i.$$

It follows easily that $(fM(\mathcal{D}_n))_{n=1}^{\infty}$ is a Cauchy sequence in $\mathbf{R}^q$. Consequently it converges to a limit A in $\mathbf{R}^q$.

To show that A is the integral of f, fix N so that $1/N < \epsilon/2$ and $|A - fM(\mathcal{D}_N)| < \epsilon/2$. Let $\mathcal{D}$ be γ_N-fine. Then

$$|fM(\mathcal{D}) - A| \leqslant |fM(\mathcal{D}) - fM(\mathcal{D}_N)| + |fM(\mathcal{D}_N) - A|$$

$$< 1/N + \epsilon/2 < \epsilon.$$

Thus $A = \int_I f$.

4. The techniques needed for the solution of this exercise anticipate the discussion on p. 75. There it is shown that $\mathcal{E}$ is a subset of a division $\mathcal{D}$ of I. Let $\mathcal{F} = \mathcal{D} - \mathcal{E}$. Then $\nu(\mathcal{E}) + \nu(\mathcal{F}) = \nu(\mathcal{D})$. Since $0 \leqslant \nu(J)$ for all J, $0 \leqslant \nu(\mathcal{F})$. Moreover $\nu(\mathcal{D}) = \nu(I)$ since ν is finitely additive. Consequently $\nu(\mathcal{E}) \leqslant \nu(I)$.

5. Let $E_t = \{x \in [a, b] : \sigma(x) = t\}$. Then $E_t \subseteq C$ when $t = \infty$ and when $t = -\infty$. When t is finite we need to show that $E_t - C$ is countable. Let $x \in E_t - C$. Then $\sigma'(x) \neq 0$, consequently there is δ_x such that $\sigma(y) \neq \sigma(x)$ when $0 < |y - x| < \delta_x$. In other words, the set $E_t - C$ is made up of isolated points. It must be countable. To see this, let $F_n = \{x \in [-n, n] : \delta_x > 1/n\}$. Then $E_t - C = \bigcup_{n=1}^{\infty} F_n$. It suffices to show that each F_n is finite. Form a division of $[-n, n]$ into subintervals of length $1/n$. Each subinterval contains at most one point of F_n. Thus F_n is finite.

6. There are gauges γ_1 and γ_2 on $[a, c]$ and $[d, b]$ such that $|\int_a^c f - fL(\mathcal{D}_1)| < \epsilon/2$ when $\mathcal{D}_1$ is a γ_1-fine division of $[a, c]$ and $|\int_d^b f - fL(\mathcal{D}_2)| < \epsilon/2$ when $\mathcal{D}_2$ is a γ_2-fine division of $[d, b]$. We may also suppose $\gamma_1(z) \subseteq \gamma(z)$ for all z in $[a, c]$ and $\gamma_2(z) \subseteq \gamma(z)$ for all z in $[d, b]$. Fix $\mathcal{D}_1$ and $\mathcal{D}_2$. Let $\mathcal{E}$ be a γ-fine division of $[c, d]$. Set $\mathcal{D} = \mathcal{D}_1 \cup \mathcal{E} \cup \mathcal{D}_2$. Then $\mathcal{D}$ is a γ-fine division of $[a, b]$. Consequently,

$$\left| \int_c^d f - fL(\mathcal{E}) \right| \leqslant \left| \int_a^b f - fL(\mathcal{D}) \right| + \left| fL(\mathcal{D}_1) - \int_a^c f \right|$$

$$+ \left| fL(\mathcal{D}_2) - \int_d^b f \right|.$$

This uses the finite additivity of the integral. Now these terms are less than ϵ, $\epsilon/2$, and $\epsilon/2$. Hence the conclusion follows.

(If $a = c$ or $b = d$, some terms become zero but otherwise the argument goes the same.)

Note: A sharpening of the argument above shows that $|\int_c^d f - fL(\mathcal{E})| \leqslant \epsilon$. This latter inequality is a special case of Henstock's lemma. (See p. 74.)

7. (a) When $f \geqslant 0$ the function $s \to \int_s^b f$ is decreasing on the interval (a, b) and nonnegative. Such a function is bounded above if and only if $\lim_{s\to a}\int_s^b f$ exists. Existence of this limit is exactly the criterion for existence of $\int_a^b f$.

(b) The function $F : s \to \int_s^b f$ has a limit at a provided the Cauchy criterion is satisfied. That is, it suffices to find c such that $|F(t) - F(s)| < \epsilon$ when $a < s < t < c$. Since $\int_a^b g$ exists, $\lim_{s\to a}\int_s^b g$ does exist. Hence there is c such that $|\int_s^b g - \int_t^b g| < \epsilon$ when $a < s < t < c$. But now

$$|F(t) - F(s)| = \left|\int_s^t f\right| \leqslant \int_s^t g = \int_s^b g - \int_t^b g < \epsilon.$$

Chapter 3

1. Since $\lim_{t\to\infty}\int_1^t 1/\sqrt{x}\ dx = \infty$, we know that $\int_1^\infty 1/\sqrt{x}\ dx$ does not exist. Set $f_n(x) = 1/\sqrt{x}$ when $1 \leqslant x \leqslant n$ and $f_n(x) = 0$ when $n < x$. Then f_n converges uniformly to $1/\sqrt{x}$ on $[1, \infty)$.

2. The function f is integrable on $[a, b]$. Set $F(c) = \lim_{n\to\infty} F_n(c)$ and $F(x) = F(c) + \int_c^x f$ elsewhere in $[a, b]$. (We are using the oriented integral here.)

We are going to get an inequality from which two separate deductions can be made. By hypothesis there is n_ϵ such that $|f_n(x) - f(x)| < \epsilon$ for all x in $[a, b]$ and all $n \geqslant n_\epsilon$. Let u and v be in $[a, b]$. Then

$$|F_n(v) - F_n(u) - (F(v) - F(u))| = \left|\int_u^v (f_n - f)\right|$$

$$\leqslant \epsilon|v - u|$$

when $n \geqslant n_\epsilon$.

The first conclusion is uniform convergence of F_n to F. To get it let $u = c$. Then one more application of the triangle inequality yields

$$|F_n(v) - F(v)| \leqslant |F_n(c) - F(c)| + \epsilon(b - a).$$

Uniform convergence follows easily.

The next conclusion is uniform convergence of difference quotients. Let $u = x$ and $v = x + t$ with $t \neq 0$. Divide by $|t|$. Then

$$\left| \frac{F_n(x + t) - F_n(x)}{t} - \frac{F(x + t) - F(x)}{t} \right| \leqslant \epsilon$$

for all x and t such that x and $x + t$ are in $[a, b]$ and all $n \geqslant n_\epsilon$.

Choose x so that $F_n'(x) = f_n(x)$ for every integer n. Select δ_ϵ such that

$$\left| \frac{F_n(x + t) - F_n(x)}{t} - f_n(x) \right| < \epsilon$$

when $n = n_\epsilon$ and $|t| < \delta_\epsilon$. Now

$$\left| \frac{F(x + t) - F(x)}{t} - f(x) \right|$$

$$\leqslant \left| \frac{F(x + t) - F(x)}{t} - \frac{F_n(x + t) - F_n(x)}{t} \right|$$

$$+ \left| \frac{F_n(x + t) - F_n(x)}{t} - f_n(x) \right| + |f_n(x) - f(x)|.$$

Each term is less than ϵ when $n \geqslant n_\epsilon$ and $|t| < \delta_\epsilon$. Consequently $F'(x) = f(x)$.

Let C_n be the set of values of x for which we cannot claim that $F_n'(x) = f_n(x)$. Then C_n is countable and $\bigcup_{n=1}^{\infty} C_n$ is also countable. Outside this set we know that

$F' = f$. Since F is continuous, it is a primitive of f on $[a, b]$.

(This argument can be greatly abbreviated by use of the iterated limits theorem given in Section S3.9.)

3. It is enough to consider increasing functions. We may apply the proposition proved in the preceding exercise. Thus it is our goal to define functions f_n which converge uniformly to f and have primitives. Step functions are the right choice. Since f is increasing, for any number K in $[f(a), f(b)]$ there is at least one number c in $[a, b]$ such that $f(x) \leqslant K$ when $x \leqslant c$ and $K \leqslant f(x)$ when $c < x$. Let the integer n be given and set $d_n = (f(b) - f(a))/n$. For $i = 1, 2, \ldots, n - 1$, let $K_i = f(a) + i d_n$. Choose a corresponding c_i. Set $c_0 = a$ and $c_n = b$. Define f_n so that $f_n(c_i) = f(c_i)$ for $i = 0, \ldots, n$ and $f_n(x) = K_i$ when $c_{i-1} < x < c_i$. It is immediate that $|f(x) - f_n(x)| < d_n$ for all x in $[a, b]$. Consequently f_n converges uniformly to f on $[a, b]$.

Chapter 4

1. Recall that E is gfM-null whenever E is fM-null. Now apply this with the constant function with value 1 in place of f and f in place of g. Thus E is fM-null when E is M-null. By definition this means that $|f|M(\mathcal{E}) < \epsilon$ for every γ-fine partial division $\mathcal{E}$ whose tags are in E. When $\mathcal{D}$ is a division of I and $\mathcal{E}$ is the subset of $\mathcal{D}$ whose tags are in E, $|f|M(\mathcal{D}) = |f|M(\mathcal{E})$ since f vanishes outside E. Thus $\int_I |f| = 0$.

2. There is a gauge γ on I such that $|f|M(\mathcal{D}) < \epsilon$ when $\mathcal{D}$ is a γ-fine division of I. Let $E = \{x \in I : f(x) \neq 0\}$. Let $\mathcal{E}$ be a γ-fine partial division of I whose tags are in E. There is a γ-fine division $\mathcal{D}$ such that $\mathcal{E} \subseteq \mathcal{D}$. Then

$|f|M(\mathcal{E}) = |f|M(\mathcal{D}) < \epsilon$. Thus E is $|f|M$-null. Let $g(x) = 1/|f(x)|$ when $x \in E$. Then E is $g|f|M$-null, i.e., M-null.

3. Let M' and M be the interval measures in $\mathbf{R}^{p-1}$ and $\mathbf{R}^p$, respectively. For a given ϵ we must define γ on E so that $M(\mathcal{E}) < \epsilon$ when $\mathcal{E}$ is a γ-fine partial division with tags in E.

Using the continuity of f choose $\gamma_1(z)$ so that $|f(x) - f(z)| < \epsilon'$ when $x \in I \cap \gamma_1(z)$. Fix a γ_1-fine division $\mathcal{D}_1$ of I. For each zJ in $\mathcal{D}_1$ let H_J be the interior of J, i.e., the open interval obtained by peeling the faces from J. Let B be the union of all faces of all J of $\mathcal{D}_1$. Then $B \times \mathbf{R}$ is an M-null subset of $\mathbf{R}^p$ since it is a finite union of degenerate intervals. Thus there is γ defined on $B \times \mathbf{R}$ such that $M(\mathcal{E}) < \epsilon/2$ when $\mathcal{E}$ is γ-fine with tags in $B \times \mathbf{R}$.

When $x \in I - B$ there is a unique J such that $x \in H_J$. Set $\gamma(x, f(x)) = H_J \times (f(z) - \epsilon', f(z) + \epsilon')$. Let $\mathcal{E}$ be γ-fine with tags in E. Separate $\mathcal{E}$ into subsets $\mathcal{E}_1$ and $\mathcal{E}_2$ having tags in $B \times \mathbf{R}$ and $(I - B) \times \mathbf{R}$. Then $M(\mathcal{E}_1) < \epsilon/2$ and $M(\mathcal{E}_2) \leqslant \sum_J M'(J) 2\epsilon' \leqslant 2\epsilon' M'(I)$. (The first inequality on $M(\mathcal{E}_2)$ results from grouping together all terms associated with a single J of $\mathcal{D}_1$.) The choice $\epsilon' = \epsilon/(4M'(I))$ gives $M(\mathcal{E}) < \epsilon$, as desired.

4. According to Example 5, p. 86, the functions $g \wedge (i^{-1}f_n)$ are integrable. There is a bound A for $\int_I f_n$. Since $g \wedge (i^{-1}f_n) \leqslant i^{-1}f_n$ for all n, $\int_I (g \wedge (i^{-1}f_n)) \leqslant i^{-1}A$. Since $f_n \leqslant f_{n+1}$ the sequence $g \wedge (i^{-1}f_n)$ increases for fixed i. Thus $\int_I h_i = \lim_{n \to \infty} \int_I (g \wedge (i^{-1}f_n)) \leqslant i^{-1}A$, too. To see that $\lim_{i \to \infty} h_i = h$, first let $x \in E$. Then $g \wedge (i^{-1}h_n)$ has value 1 at x for all large n. Hence $h_i(x) = 1$ and $\lim_{i \to \infty} h_i(x) = 1$. When $x \in I - E$, $(g \wedge (i^{-1}f_n))(x) \leqslant i^{-1}f(x)$, hence $h_i(x) \leqslant i^{-1}f(x)$. Consequently, $\lim_{i \to \infty} h_i(x) = 0$. A second use of monotone convergence gives $\int_I h = \lim_{i \to \infty} \int_I h_i = 0$.

5. The essentials of the solution can be conveyed by an example in $\mathbf{N}$. Consider the sequence $1, -1, 1, 1/2, -1/2, 1/2, 1/3, -1/3, 1/3, \ldots$. This is our function $\cdot f : \mathbf{N} \to \mathbf{R}$. Note that the terms go in groups of three. Let's replace every third member by zero. That means integrate f over $E = \{1, 2, 4, 5, 7, 8, \ldots\}$. It is easy to see that $\int_E f$ exists. Next replace the first member of each group by zero; i.e., integrate f over $F = \{2, 3, 5, 6, 8, 9, \ldots\}$. Integration over $E \cap F$ is the same as summing the middle terms of each group. These terms are negatives of terms of the harmonic series. Thus $\int_{E \cap F} f$ does not exist. Neither does $\int_{E \cup F} f$ for much the same reason.

6. For each positive integer n, partition I into a finite collection $\mathcal{F}_n$ of pairwise disjoint intervals K such that the diameter of K is no more than $1/n$. Also make the choices so that each interval of $\mathcal{F}_{n+1}$ is contained in an interval of $\mathcal{F}_n$; i.e., $\mathcal{F}_{n+1}$ is a refinement of $\mathcal{F}_n$.

Define f_n from $\mathcal{F}_n$ as follows. For each x in I, there is just one interval K of $\mathcal{F}_n$ to which x belongs. Let $f_n(x) = \text{lub}\{f(x) : x \in K\}$. This is meaningful everywhere since f is bounded above. (Recall that f has a maximum on E and f is constant on $I - E$.)

The sequence $(f_n)_{n=1}^{\infty}$ decreases since $\mathcal{F}_{n+1}$ is a refinement of $\mathcal{F}_n$.

Now let $x \in I - E$. There is an open interval $\gamma(x)$ which does not intersect E since E is closed. Every interval of sufficiently small diameter containing x is a subset of $\gamma(x)$. Since f vanishes on $\gamma(x)$ so does $f_n(x) = 0$ for sufficiently large n.

When $x \in E$ there is an open interval $\gamma(x)$ such that $|f(y) - f(x)| < \epsilon$ for all $y \in E \cap \gamma(x)$. There may be points in $\gamma(x)$ which are not in E. But f is zero at those points. Since $0 \leqslant f(x)$, $f(y) \leqslant f(x) + \epsilon$ for all y in $\gamma(x)$. For large enough n the interval of $\mathcal{F}_n$ which contains x is a subset of $\gamma(x)$. Thus $f_n(x) \leqslant f(x) + \epsilon$ for all sufficiently large n. The definition of f_n also insures that $f(x) \leqslant f_n(x)$. Hence $f(x) \leqslant f_n(x) \leqslant f(x) + \epsilon$ for sufficiently large n.

We have shown that $f(x) = \lim_{n \to \infty} f_n(x)$ for all $x \in I$, as required.

7. Set $E_1 = H_1$ and $E_n = H_n - H_{n-1}$ for $n \geqslant 2$. Note that $H_{n-1} \subseteq H_n$. Thus f is integrable on E_n. Since f is also bounded on each E_n, it is absolutely integrable on E_n. Moreover $\int_{E_n} |f| \leqslant b_n(M(H_n) - M(H_{n-1}))$ for $n \geqslant 2$. But $M(H_n) - M(H_{n-1}) = (2n)^p - (2(n-1))^p \leqslant p2^p n^{p-1}$. Thus $\sum_{n=1}^{\infty} \int_{E_n} |f|$ is convergent since $\sum_{n=1}^{\infty} b_n n^{p-1}$ is assumed convergent. This implies absolute integrability of f.

8. We know that $|f|$ is integrable on E_n for all n since $|f| \leqslant g$ and g is integrable on I. By induction f is absolutely integrable on $\bigcup_{i=1}^{n} E_i$ for all n since absolute integrability carries over to unions of two sets. Let $F_1 = E_1$ and $F_n = \bigcup_{i=1}^{n} E_i - (\bigcup_{i=1}^{n-1} E_i)$ when $n \geqslant 2$. Now f is absolutely integrable on each member of the pairwise disjoint sequence $(F_n)_{n=1}^{\infty}$. Moreover $\sum_{i=1}^{n} \int_{F_n} |f| \leqslant \int_I g$ for all n. Thus $\sum_{n=1}^{\infty} \int_{F_n} |f|$ is finite. By the countable additivity proposition f is absolutely integrable on $\bigcup_{n=1}^{\infty} F_n$, i.e., on E.

9. The characteristic functions of $\bigcap_{i=1}^{n} E_i$ decrease to the characteristic function of $\bigcap_{i=1}^{\infty} E_i$. Thus the monotone convergence theorem yields the integrability of the intersection of the sequence.

Now $\bigcup_{i=1}^{n} E_i$ are integrable sets whose characteristic functions increase to the characteristic function of the union of the sequence. Since $\mu(\bigcup_{i=1}^{n} E_i) \leqslant \sum_{i=1}^{n} \mu(E_i)$ and $\mu(\bigcup_{i=1}^{n} E_i) \leqslant \mu(F)$ when $\bigcup_{i=1}^{\infty} E_i$ is contained in the integrable set F, the monotone convergence theorem implies integrability of $\bigcup_{i=1}^{\infty} E_i$ under each of the conditions which have been given.

10. The conclusions in (a) follow from known properties of integrable sets, since the intersection with a bounded interval J distributes over the other operations. For instance, $(E \cup F) \cap J = (E \cap J) \cup (F \cap J)$. Thus $(E \cup F) \cap J$ is the union of integrable sets and is integrable itself. The others go similarly.

Measurability of $\bigcap_{n=1}^{\infty} E_n$ and $\bigcup_{n=1}^{\infty} E_n$ follow similarly. The inequality is trivial when the right-hand side equals ∞. When it is finite, every set E_n is integrable and we know from the previous exercise that $\bigcup_{n=1}^{\infty} E_n$ is integrable. Moreover $\mu(\bigcup_{i=1}^{n} E_i) \leqslant \sum_{i=1}^{n} \mu(E_i) \leqslant \sum_{i=1}^{\infty} \mu(E_i)$ for all n. Monotone convergence tells us that $\mu(\bigcup_{i=1}^{\infty} E_i) \leqslant \sum_{i=1}^{\infty} \mu(E_i)$.

When the sets are nonoverlapping and $\bigcup_{n=1}^{\infty} E_n$ is not integrable the inequality just proved becomes equality. When $\bigcup_{n=1}^{\infty} E_n$ is integrable, we are back to countable additivity of integrable sets since every E_n must also be integrable.

Chapter 5

1. We prove that (i) implies (ii) implies (iii) implies (iv) implies (i).

Assume (i). Let $G = (-\infty, a]$ and $G' = (a, \infty)$. Then $f^{-1}(G) = I - f^{-1}(G')$. Since $f^{-1}(G')$ is measurable according to (i) the set $f^{-1}(G)$ is also measurable. This establishes (ii).

Assume (ii). Let $G = (-\infty, a)$ and set $G_n = (-\infty, a - 1/n]$. Then $\bigcup_{n=1}^{\infty} G_n = G$ and $f^{-1}(G) = \bigcup_{n=1}^{\infty} f^{-1}(G_n)$. By (ii) each $f^{-1}(G_n)$ is measurable. Thus $f^{-1}(G)$ is also measurable and (iii) holds.

Prove (iv) from (iii) like (ii) from (i).

Deduce (i) from (iv) on the model of (iii) from (ii).

Measurability of f obviously implies (i), hence all the others. Assume (i) through (iv). Then $f^{-1}(G)$ is measurable for every unbounded open interval. Since

$(a, b) = (a, \infty) - [b, \infty)$ and $f^{-1}((a, b)) = f^{-1}((a, \infty)) - f^{-1}([b, \infty))$, it follows from (i) and (ii) that $f^{-1}(G)$ is measurable for bounded open intervals, too. Thus f is measurable.

2. In $\mathbf{R}^q$ let $J_n = [-n, n) \times \cdots \times [-n, n)$. Partition each factor $[-n, n)$ into $n2^n$ intervals of length $1/2^{n-1}$. Let each subinterval be closed on the left and open on the right. Form a partition $\mathcal{F}_n$ of J_n by taking Cartesian products. (See Fig. 3, p. 144.)

For each $G \in \mathcal{F}_n$ the set $I_n \cap f^{-1}(G)$ is a bounded measurable subset of I. Let K be the closed interval having the same faces as G. Then K has on its boundary a unique point y which is nearest to the origin. Set $f_n(x) = y$ for all x in $I_n \cap f^{-1}(G)$. Do this for all G in $\mathcal{F}_n$. This defines f_n on $I_n \cap f^{-1}(J_n)$. Set $f_n(x) = 0$ elsewhere. Then $|f_n(x)| \leqslant |f(x)|$ for all x in I. Since $\mathcal{F}_{n+1}$ is a refinement of $\mathcal{F}_n$ and $I_n \subseteq I_{n+1}$, $|f_n(x)| \leqslant |f_{n+1}(x)|$ for all x. Moreover this construction produces a nonnegative function f_n when f is nonnegative.

The convergence of $f_n(x)$ to $f(x)$ is proved as follows. Let $x \in I \cap \mathbf{R}^p$. There is n_x such that $x \in I_n$ and $f(x) \in J_n$ when $n \geqslant n_x$. Then $f(x)$ is in some G of $\mathcal{F}_n$. Since $f_n(x)$ belongs to the closed interval having the same faces as G, $|f(x) - f_n(x)| \leqslant \sqrt{q}/2^{n-1}$. Consequently, $\lim_{n \to \infty} f_n(x) = f(x)$.

3. Let $E = \bigcup_{n=1}^{\infty} \bigcup_{m=1}^{\infty} \bigcap_{k=m}^{\infty} f_k^{-1}(G_n)$. Let $x \in E$. Then $x \in \bigcap_{k=m}^{\infty} f_k^{-1}(G_n)$ for some m and n. Consequently, $f_k(x) > a + 1/n$ when $k \geqslant m$. Thus $\lim_{k \to \infty} f_k(x) \geqslant a + 1/n$. From this we see that $x \in f^{-1}(G)$. Thus $E \subseteq f^{-1}(G)$.

Conversely, let $x \in f^{-1}(G)$. Fix n so that $a + 1/n < f(x)$. There is m such that $f_k(x) > a + 1/n$ when $k \geqslant m$. Then $x \in f_k^{-1}(G)$ when $k \geqslant m$; i.e., $x \in \bigcap_{k=m}^{\infty} f_k^{-1}(G_n)$. Consequently, $x \in E$. This completes the proof that $f^{-1}(G) \subseteq E$. Thus the sets are equal.

4. If one of f and g is a null function the same is true of fg. Then both sides of the inequality are zero.

Suppose neither f nor g is a null function. Then $\int_I |f|^s > 0$ and $\int_I |g|^t > 0$. We can choose a positive constant c so that $\int_I |cf|^s = 1$. In fact we need only have $c^{-s} = \int_I |f|^s$ or $c^{-1} = (\int_I |f|^s)^{1/s}$. Similarly, $\int_I |kg|^t = 1$ when $k^{-1} = (\int_I |g|^t)^{1/t}$. Integration on both sides of

$$|(cf)(kg)| \leqslant |cf|^s/s + |kg|^t/t$$

yields

$$ck \int_I |fg| \leqslant 1/s + 1/t = 1.$$

Hence $\int_I |fg| \leqslant c^{-1}k^{-1}$, as desired.

Chapter 6

1. Some notation is needed. Suppose $I \subseteq \overline{\mathbf{R}}^p$. Let the r-fold integration be accomplished by expressing $\overline{\mathbf{R}}^p$ as $P_1 \times P_2 \times \cdots \times P_r$ with $r \leqslant p$. Write the intervals as products like this: $I = I_1 \times I_2 \times \cdots \times I_r$, where $I_j \subseteq P_j$ for $1 \leqslant j \leqslant r$.

We may assume $I = G \cup H$. Since G and H do not overlap, there is an integer k such that $I_j = G_j = H_j$ when $j \neq k$ and $G_k \cup H_k = I_k$ with G_k and H_k nonoverlapping.

For convenience denote iterated integrals as follows. When $i < r$ let the result of the first i integrations over an interval J be

$$\nu(J; x_{i+1}, \ldots, x_r)$$

$$= \int_{J_i} \cdots \int_{J_2} \int_{J_1} f(x_1, x_2, \ldots, x_r) \, dx_1 \, dx_2 \cdots dx_i.$$

Let $\nu(J)$ be the result of the r-fold integration.

The equalities $I_j = G_j = H_j$ when $j < k$ imply

$$\nu(I; x_{i+1}, \ldots, x_r) = \nu(G; x_{i+1}, \ldots, x_r)$$
$$= \nu(H; x_{i+1}, \ldots, x_r)$$

when $i < k$. The additivity of integrals allows us to replace the second equality by summation when $i = k$. The result is either

$$\nu(I; x_{k+1}, \ldots, x_r) = \nu(G; x_{k+1}, \ldots, x_r)$$
$$+ \nu(H; x_{k+1}, \ldots, x_r)$$

when $k < r$ or, of course, $\nu(I) = \nu(G) + \nu(H)$ if it happens that $k = r$. When $k < r$ the linearity of integration applies in the last $r - k$ integrations to yield the desired final conclusion that $\nu(I) = \nu(G) + \nu(H)$. The equalities $I_j = G_j = H_j$ for $j = k + 1, \ldots, r$ must be used in these last $r - k$ integrations.

2. Since U can be expressed as a countable union of bounded sets, it is enough to give the solution for a bounded interval I. It is enough to show that there is a gauge γ on U such that

$$|f(z)M(J) - \phi(J)| < \epsilon M(J)$$

when $z \in J$ and $J \subseteq \gamma(z)$.

When $z \in U$, the continuity of f at z implies that there is $\gamma(z)$ such that $|f(x) - f(z)| < \epsilon$ when $x \in I \cap \gamma(z)$. Let $J \subseteq I \cap \gamma(z)$. On J consider the constant function g such that $g(x) = f(z)$. The iterated integral of g over J has the value $f(z)M(J)$. Then $\phi(J) - f(z)M(J)$ is the iterated integral of $f - g$. Since $|f(x) - g(x)| < \epsilon$ on J, repeated application of integral inequalities yields $|\phi(J) - f(z) \cdot M(J)| < \epsilon M(J)$.

3. The first step is to snow that the extension of f which is zero outside E has an iterated integral over any bounded interval. Actually, it is enough to work within $I = [a, b] \times [c, d]$ where $c \leqslant g(x) \leqslant h(x) \leqslant d$ for all x in $[a, b]$. The order of integration will always be the same as this:

$$\phi(I) = \int_a^b \int_c^d f(x, y) \, dy \, dx.$$

The inner integral exists because, for fixed x, $f(x, y)$ vanishes on $[c, g(x))$ and $(h(x), d]$ and is continuous when restricted to $[g(x), h(x)]$.

Let $[r, s] \subseteq [c, d]$. The outer integral exists if we show $\int_r^s f(x, y) \, dy$ is a continuous function of x. Thus it is appropriate to consider $\int_r^s f(x, y) \, dy - \int_r^s f(u, y) \, dy$, i.e., $\int_r^s [f(x, y) - f(u, y)] \, dy$. It is important to estimate this integrand.

Let $K_x = [r, s] \cap [g(x), h(x)]$ for each x in $[a, b]$. The uniform continuity of f on E implies that $|f(x, y) - f(u, y)| < \epsilon$ when $|x - u| < \delta$ and $y \in K_x \cap K_u$. Since f is bounded, there is a constant A such that $|f(x, y) - f(u, y)| \leqslant A$ for all x, u, and y. Next note that $f(x, y) - f(u, y) = 0$ when y is outside $K_x \cup K_u$. Finally, note that the total length of the intervals making up $(K_x \cup K_y) - (K_x \cap K_y)$ is no more than $|g(x) - g(u)| + |h(x) - h(u)|$.

Now suppose $|x - u| < \delta$. The estimates of the preceding paragraph give

$$\left| \int_r^s [f(x, y) - f(u, y)] \, dy \right|$$
$$\leqslant \epsilon L(K_x \cap K_u) + A(|g(x) - g(u)| + |h(x) - h(u)|).$$

Since $L(K_x \cap K_u) \leqslant d - c$ and g and h are continuous, the desired continuity of the inner integral follows.

Now $\phi(J)$ is meaningful on every J contained in I. Let G and H be the graphs of g and h. Then f is continuous on $I - (G \cup H)$ and this set is $(fM - \phi)$-null. We know $G \cup H$ is M-null. It is also ϕ-null since $|\phi(J)| \leqslant AM(J)$ for every J. It follows that f is integrable on E and $\int_E f = \phi(I)$.

Chapter 7

1. Let A be the $\mathscr{N}$-limit of $f\Delta\alpha(\mathscr{D})$. Choose δ so that $|A - f\Delta\alpha(\mathscr{D})| < \epsilon$ when $\|\mathscr{D}\| < \delta$. Fix a division $\mathscr{F}$ with $\|\mathscr{F}\| < \delta$. Let $\mathscr{D}$ be a refinement of $\mathscr{F}$. Then $\|\mathscr{D}\| \leqslant \|\mathscr{F}\| < \delta$. Hence $|A - f\Delta\alpha(\mathscr{D})| < \epsilon$. Therefore A is also the $\mathscr{R}$-limit of $f\Delta\alpha(\mathscr{D})$.

Now assume A is the $\mathscr{R}$-limit. To show that A is the $\mathscr{G}$-limit requires the use of a special gauge associated with a division. Fix a division $\mathscr{F}$ for which $|A - f\Delta\alpha(\mathscr{D})| < \epsilon$ when $\mathscr{D}$ is a refinement of $\mathscr{F}$. Let $\gamma_{\mathscr{F}}$ be defined so that $\gamma_{\mathscr{F}}(z)$ contains no endpoint of $\mathscr{F}$ distinct from z. Let $\mathscr{D}$ be $\gamma_{\mathscr{F}}$-fine. Then every endpoint of $\mathscr{F}$ appears as tag of each interval of $\mathscr{D}$ which contains that endpoint. Form $\mathscr{E}$ from $\mathscr{D}$ by replacing $z[u, v]$ by $z[u, z]$ and $z[z, v]$ when $u < z < v$. Then $\mathscr{E}$ is a refinement of $\mathscr{F}$. Moreover $f\Delta\alpha(\mathscr{D}) = f\Delta\alpha(\mathscr{E})$. Thus $|A - f\Delta\alpha(\mathscr{D})| = |A - f\Delta\alpha(\mathscr{E})| < \epsilon$ because of the choice of $\mathscr{F}$.

2. It is enough to find γ for which $|f\Delta\alpha(\mathscr{D}) - fg\Delta\beta(\mathscr{D})| \leqslant \epsilon$ for all γ-fine $\mathscr{D}$. From this statement and the triangle inequality we can deduce existence of both integrals from existence of either of them.

Let $E_n = \{x \in [a, b] : n - 1 \leqslant |f(x)| < n\}$. These sets E_n cover $[a, b]$ and are pairwise disjoint. Any division $\mathscr{D}$ falls into subsets $\mathscr{D}_n$ having tags in E_n. Now $|f\Delta\alpha(\mathscr{D}_n) - fg\Delta\beta(\mathscr{D}_n)| \leqslant n|\Delta\alpha - g\Delta\beta|(\mathscr{D}_n)$. Thus it is our goal to define γ so that $\sum_n n|\Delta\alpha - g\Delta\beta|(\mathscr{D}_n) < \epsilon$ when $\mathscr{D}$ is γ-fine. Henstock's lemma gets us to our goal.

Begin with gauges γ_n on $[a, b]$ so that $|\int_a^b g d\beta - g\Delta\beta(\mathcal{D})|$ $< \epsilon/(nq2^{n+1})$ when $\mathcal{D}$ is γ_n-fine. When $z \in E_n$ set $\gamma(z) = \gamma_n(z)$. Then $\mathcal{D}_n$ is γ_n-fine provided $\mathcal{D}$ is γ-fine. Thus $|\Delta\alpha - g\Delta\beta|(\mathcal{D}_n) \leq 2q\epsilon/(nq2^{n+1})$ and $|f\Delta\alpha(\mathcal{D}_n) - fg\Delta\beta(\mathcal{D}_n)| \leq \epsilon/2^n$. Summation on n does the rest.

3. The step function f can be expressed as a linear combination of functions of the types considered in Example 3, p. 190, and Example 4, p. 190. The n open intervals give us functions g_j such that $g_j(x) = 1$ when $x \in (x_{j-1}, x_j)$ and $g_j(x) = 0$ elsewhere. The $n + 1$ endpoints give us functions h_j such that $h_j(x_j) = 1$ and $h_j(x) = 0$ elsewhere. Then $f = \sum_{j=1}^n F_j g_j + \sum_{j=0}^n f(x_j) h_j$.

Part (a) follows immediately from linearity in the integrand and the results given in Examples 3 and 4.

Part (b) can be done most easily from the definition. When $\mathcal{D}$ has every x_j as a tag and at least one endpoint in each interval (x_{j-1}, x_j), the value of $\alpha\Delta f(\mathcal{D})$ is precisely the expression given in (b). It suffices to define γ so that $\gamma(z)$ contains no x_j distinct from z. Then any γ-fine $\mathcal{D}$ has the properties named and the conclusion is immediate.

4. The jumps in f occur on the left-hand side of the integer points. Consequently, from Exercise 3(b), $\int_0^t \alpha \, df$ $= \sum_{j=1}^n \alpha(x_j)$ when $n \leq t < n + 1$. Now $\int_0^\infty \alpha \, df$ exists if and only if $\lim_{t\to\infty} \int_0^t \alpha \, df$ exists. Consequently, the existence of the integral is equivalent to convergence of $\sum_{j=1}^\infty \alpha(x_j)$. Moreover $\int_0^\infty \alpha \, df = \sum_{j=1}^\infty \alpha(x_j)$.

5. Let α be the primitive F of Example 2, p. 79. Since this function is continuous, it is a regulated function. It was constructed so that it is not a function of bounded variation.

Since α is a primitive of α', $\alpha^2/2$ is also a primitive of $\alpha\alpha'$ and $\int_a^b \alpha\alpha'$ exists. It can be converted into $\int_a^b \alpha \, d\alpha$.

The failure of $\int_a^b f \, d\alpha$ to exist is the same as failure of existence of $\int_a^b f\alpha'$. Clearly, it is desirable that $f(x)$ and

$\alpha'(x)$ have the same sign. The sign of $\alpha'(x)$ is alternately positive and negative in intervals (c_0, c_1), (c_1, c_2), ..., beginning with a positive value in (c_0, c_1). The value of α' on (c_{n-1}, c_n) is $a_n/(c_n - c_{n-1})$ where $(a_n)_{n=1}^{\infty}$ is the sequence $1, -1/2, 1/2, -1/3, 1/3, \ldots$. On $[c_{n-1}, c_n)$ let $f(x) = b_n$ where $(b_n)_{n=1}^{\infty}$ is the sequence $1, -1/\ln 2, 1/\ln 2, -1/\ln 3, 1/\ln 3, \ldots$. Then $\sum_{n=1}^{\infty} a_n b_n$ is divergent. Since $\int_a^{c_n} f\alpha' = \sum_{k=1}^{n} a_k b_k$, it follows that $\lim_{t \to b} \int_a^t f\alpha'$ does not exist. Consequently, $\int_a^b f\alpha'$ and $\int_a^b f \, d\alpha$ do not exist. Finally, observe that f is regulated. It clearly has one-sided limits at each point in $[a, b]$. Since $\lim_{n \to \infty} b_n = 0$ the left-hand limit of f at b is also zero.

6. Let $C = \{c_1, c_2, c_3, \ldots\}$ include all left-hand discontinuities of α, whether $\phi(c_n) \neq 0$ or not. Then $\sum_{n=1}^{\infty} \phi(c_n)$ is absolutely convergent and there is m such that $\sum_{k=m+1}^{\infty} |\phi(c_k)| < \epsilon$. (We may assume $a \notin C$.)

When $z = c_n$ choose $\gamma(z)$ so that

$$|\phi(z) - (\alpha(z) - \alpha(u))(f(z) - f(u))| < \epsilon/2^n$$

for all u such that $u < z$ and $u \in \gamma(z)$.

When $z \notin C$ the function α is left-hand continuous at z. Thus there is $\gamma(z)$ such that

$$|(\alpha(z) - \alpha(u))(f(z) - f(u))| \leqslant \epsilon |f(z) - f(u)|$$

when $u \leqslant z$ and $u \in \gamma(z)$.

When $z \notin \{c_1, c_2, \ldots, c_m\}$ restrict $\gamma(z)$ further so that $\gamma(z)$ contains none of $c_1, c_2, \ldots, c_m$.

Let $\mathcal{D}$ be γ-fine. Break $\mathcal{D}$ into three subsets. Put $z[u, v]$ into $\mathcal{E}$ when $z \notin C$, into $\mathcal{F}$ when $z \in C$ and $u < z$, and into $\mathcal{G}$ when $z \in C$ but $u = z$. Let K be the set of all n such that c_n is a tag in $\mathcal{F}$. Then $\{1, 2, \ldots, m\} \subseteq K$. Recall that the third restriction on γ implies that each interval of $\mathcal{D}$ which contains c_n, $n \leqslant m$, has c_n as its tag. Whether $\mathcal{D}$

contains one or two intervals to which c_n belongs, one of them does not have c_n as its left endpoint since $a \neq c_n$.

The definition of Φ implies that $\Phi(\mathcal{G}) = 0$. Thus

$$\left| \Phi(\mathcal{D}) - \sum_{n=1}^{\infty} \phi(c_n) \right| \leqslant \left| \Phi(\mathcal{F}) - \sum_{n \in K} \phi(c_n) \right| + |\Phi(\mathcal{E})|$$
$$+ \left| \sum_{n \notin K} \phi(c_n) \right|.$$

For each $n \in K$ there is exactly one $z[u, v]$ of $\mathcal{F}$ with $z = c_n$. Thus $|\Phi(\mathcal{F}) - \sum_{n \in K} \phi(c_n)| \leqslant \sum_{n \in K} \epsilon/2^n < \epsilon$. Recall that f is a function of bounded variation. The choice of $\gamma(z)$ when $z \notin C$ allows us to say that $|\Phi(\mathcal{E})| \leqslant \epsilon |\Delta f|(\mathcal{E}) \leqslant \epsilon V_a^b f$. Finally $|\sum_{n \notin K} \phi(c_n)| \leqslant \sum_{n=m+1}^{\infty} |\phi(c_n)| < \epsilon$. In summary

$$\left| \Phi(\mathcal{D}) - \sum_{n=1}^{\infty} \phi(c_n) \right| < \epsilon (2 + V_a^b f).$$

One more observation is needed to complete the proof. The set C was chosen to include all points where α is not left-hand continuous. The sum of the series $\sum_{n=1}^{\infty} \phi(c_n)$ is the same if some other sequence $(c_n)_{n=1}^{\infty}$ is used so long as all points where ϕ is nonzero are included. Thus the proposition is true as stated.

7. Suppose F has bounded variation on $[a, b]$. According to p. 223 the variation of F satisfies

$$V_a^b F = \lim_{t \to b} V_a^t F + \lim_{t \to b} |F(b) - F(t)|.$$

Since α has bounded variation on $[a, t]$, we already know that $V_a^t F = \int_a^t |f(x)| \, dV_a^x \alpha$. Moreover, $\lim_{t \to b} |F(b) - F(t)| = \lim_{t \to b} |\int_t^b f \, d\alpha|$. But from p. 187 we see that this last limit is zero since $\Delta\alpha([t, b]) = 0$. From the same source $\int_a^b |f(x)| \, dV_a^x \alpha = \lim_{t \to b} \int_a^t |f(x)| \, dV_a^x \alpha$ since the variation of α, like α, is defined on $[a, b)$. Now $\int_a^b |f(x)| \, dV_a^x \alpha = V_a^b F$, as required.

8. We know that $\int_c^v f \circ \tau \, d(\alpha \circ \tau) = \int_{\tau(c)}^{\tau(v)} f \, d\alpha$ when $v < d$. For convenience set $g = f \circ \tau$ and $\beta = \alpha \circ \tau$. Then β is continuous at d and $\lim_{v \to d} \Delta \beta([v, \ d]) = 0$. Consequently, existence of $\int_c^d g \, d\beta$ is equivalent to existence of $\lim_{v \to d} \int_c^v g \, d\beta$ and $\int_c^d g \, d\beta = \lim_{v \to d} \int_c^v g \, d\beta$. The existence of this limit can be determined by examining $\lim_{v \to d} \int_{\tau(c)}^{\tau(v)} f \, d\alpha$. Set $F(x) = \int_{\tau(c)}^x f \, d\alpha$. Then F is continuous at $x = \tau(d)$ since α is continuous there. The next point to note is that $F \circ \tau$ is continuous at d. Thus

$$\int_{\tau(c)}^{\tau(d)} f \, d\alpha = \lim_{v \to d} \int_{\tau(c)}^{\tau(v)} f \, d\alpha = \lim_{v \to d} \int_c^v g \, d\beta = \int_c^d g \, d\beta.$$

9. Let $a = b_1 < c_1 < b_2 < c_2 < \cdots$ with $\lim_{n \to \infty} b_n = b$. For every n let $\alpha(b_n) = 0$ and $\alpha(c_n) = 2/\sqrt{n}$. Let α be linear on the intervals $[b_n, c_n]$ and $[c_n, b_{n+1}]$. Finally, let $\alpha(b) = 0$.

On the intervals $[b_n, c_n]$ where α is increasing we will assign f positive values and on $[c_n, b_{n+1}]$ where α is decreasing f will be negative. To be specific, let $f(b_n) = f(c_n) = 0$ for all n. At the midpoint of $[b_n, c_n]$ let f have the value $1/\sqrt{n}$. Midway between c_n and b_{n+1} let f have the value $-1/\sqrt{n}$. Between these points let f be linear. Let $f(b) = 0$. Then f is continuous on $[a, b]$. Moreover $\int_{b_n}^{c_n} f \, d\alpha = \int_{b_n}^{c_n} f \alpha'$. On (b_n, c_n) the derivative α' is constant with value $\alpha(c_n)/(c_n - b_n)$. Moreover $\int_{b_n}^{c_n} f$ is the area of a triangle with altitude $1/\sqrt{n}$ and base $c_n - b_n$. Thus $\int_{b_n}^{c_n} f \, d\alpha = 1/n$. A similar analysis shows that $\int_{c_n}^{b_{n+1}} f \, d\alpha = 1/n$.

Let $\alpha_n(x) = \alpha(x)$ when $0 \leqslant x \leqslant b_n$ and $\alpha_n(x) = 0$ when $b_n < x \leqslant b$. Then α_n converges uniformly to α on $[a, b]$. Moreover $\int_a^b f \, d\alpha_n = \int_a^{b_n} f \, d\alpha = \sum_{k=1}^{n-1} 2/k$. Consequently, $\lim_{n \to \infty} \int_a^b f \, d\alpha_n$ does not exist.

INDEX